T0305506

Technology Innovation Pillars for Industry 4.0

Technology Innovation Pillars for Industry 4.0: Challenges, Improvements, and Case Studies discusses the latest innovations in the application of technologies to Industry 4.0 and the nine pillars and how they relate, support, and bridge the gap between the digital and physical worlds we now live in.

This book discusses each of the nine pillars and the roles they play in the rapid transformation of the design and operation, and offers applications and case studies supporting Industry 4.0 technologies. It presents the supply chain organizational activities utilizing cyber-physical systems architectures and talks about the advantages of intelligent manufacturing and the ability to proactively detect and respond to events, to improve quality and yield, reduce downtime, and lead to better overall equipment effectiveness among other advantages in smart factory operations.

This reference book provides a great resource for undergraduate and graduate students, industrial and manufacturing engineers, and engineers of related disciplines along with business professionals, explaining what the nine pillars are and how they relate to Industry 4.0 and smart factories.

Intelligent Manufacturing and Industrial Engineering

Series Editor: Ahmed A. Elngar, Beni-Suef University, Egypt
Mohamed Elhoseny, Mansoura University, Egypt

Machine Learning Adoption in Blockchain-Based Intelligent Manufacturing
Edited by Om Prakash Jena, Sabyasachi Pramanik, Ahmed A. Elngar

Integration of AI-Based Manufacturing and Industrial Engineering Systems with the Internet of Things
Edited by Pankaj Bhambri, Sita Rani, Valentina E. Balas and Ahmed A. Elngar

AI-Driven Digital Twin and Industry 4.0: A Conceptual Framework with Applications
Edited by Sita Rani, Pankaj Bhambri, Sachin Kumar, Piyush Kumar Pareek, and Ahmed A. Elngar

Technology Innovation Pillars for Industry 4.0: Challenges, Improvements, and Case Studies
Edited by Ahmed A. Elngar, N. Thillaiarasu, T. Saravanan, and Valentina Emilia Balas

For more information about this series, please visit: https://www.routledge.com/Mathematical-Engineering-Manufacturing-and-Management-Sciences/book-series/CRCIMIE

Technology Innovation Pillars for Industry 4.0

Challenges, Improvements, and Case Studies

Edited by
Ahmed A. Elngar, N. Thillaiarasu, T. Saravanan,
and Valentina Emilia Balas

CRC Press
Taylor & Francis Group
Boca Raton London New York

CRC Press is an imprint of the
Taylor & Francis Group, an **informa** business

Designed cover image: Shutterstock - Chor muang

First edition published 2025
by CRC Press
2385 NW Executive Center Drive, Suite 320, Boca Raton FL 33431

and by CRC Press
4 Park Square, Milton Park, Abingdon, Oxon, OX14 4RN

CRC Press is an imprint of Taylor & Francis Group, LLC

© 2025 selection and editorial matter, Ahmed A. Elngar, N. Thillaiarasu, T. Saravanan, and Valentina Emilia Balas; individual chapters, the contributors

Reasonable efforts have been made to publish reliable data and information, but the author and publisher cannot assume responsibility for the validity of all materials or the consequences of their use. The authors and publishers have attempted to trace the copyright holders of all material reproduced in this publication and apologize to copyright holders if permission to publish in this form has not been obtained. If any copyright material has not been acknowledged please write and let us know so we may rectify in any future reprint.

Except as permitted under U.S. Copyright Law, no part of this book may be reprinted, reproduced, transmitted, or utilized in any form by any electronic, mechanical, or other means, now known or hereafter invented, including photocopying, microfilming, and recording, or in any information storage or retrieval system, without written permission from the publishers.

For permission to photocopy or use material electronically from this work, access www.copyright.com or contact the Copyright Clearance Center, Inc. (CCC), 222 Rosewood Drive, Danvers, MA 01923, 978-750-8400. For works that are not available on CCC please contact mpkbookspermissions@tandf.co.uk

Trademark notice: Product or corporate names may be trademarks or registered trademarks and are used only for identification and explanation without intent to infringe.

ISBN: 978-1-032-47839-5 (hbk)
ISBN: 978-1-032-48276-7 (pbk)
ISBN: 978-1-003-38824-1 (ebk)

DOI: 10.1201/9781003388241

Typeset in Times
by Newgen Publishing UK

Contents

ARTIFICIAL INTELLIGENCE

BIG DATA ANALYTICS

CLOUD AND SECURITY

IOT

DIGITIZATION OF INDUSTRIAL PROCESSES

INFORMATION SYSTEM IN INDUSTRY

ADDITIVE MANUFACTURING

Preface

This book, entitled *Technology Innovation Pillars for Industry 4.0: Challenges, Improvements, and Case Studies*, consists of ten chapters. The book focuses on rapid transformations in the design and operations of the Internet of Things, cloud computing, autonomous and robotic systems, big data analytics, augmented reality, cyber security, simulation, system integration, and additive manufacturing. Industry 4.0 principles are already changing the way that companies deal with their daily tasks. Many aspects of smart factories can use the technologies promoted by Industry 4.0. Intelligent manufacturing is a broad concept of manufacturing that makes full use of advanced information and manufacturing technologies to optimize production and product transactions, improve quality, reduce downtime, and improve overall equipment effectiveness.

This book provides a forum for scientists, researchers, students, and practitioners to present their latest research results, ideas, developments, and applications. The book is organized into ten chapters, which include explanations of topics and relevant algorithms.

Chapter 1 explores the fundamental Role of Artificial Intelligence in Telecommunication Systems: A Healthcare Perspective.

Chapter 2 discusses An Intelligent System Utilizing Bipolar Fuzzy Logic for Ensuring Semantic Interoperability and Privacy Preservation in Healthcare Systems.

Chapter 3 includes Graph Optimizations in Neural Networks by ONNX Model.

Chapter 4 discusses Convolutional Neural Network Architecture for Accurate Plant Classification.

Chapter 5 looks at Big Data Visualizing with Augmented and Virtual Reality: Challenges and Research Agenda.

Chapter 6 discusses Mathematical Model for Service-Selection Optimization and Scheduling in Cloud Manufacturing Using Sub-Task Scheduling with Fuzzy Inference Rule.

Chapter 7 examines Social Media Initiatives through IoT to Link the Bridge between Industrial Demands with Higher Education Millennial Students through Experience Learning.

Chapter 8 discusses Analyzing Consumer Product Feedback Dynamics with Confidence Intervals.

Chapter 9 addresses Amplifying the Effectiveness of a Learning Management System: Exploring the Impact of NEP-Compliant Curriculum Changes on Higher Education Institutions.

Chapter 10 details The Future of Immersive Experience: Exploring Metaverse Application Development Technologies and Tools.

About the Editors

Ahmed A. Elngar is the founder and head of Scientific Innovation Research Group (SIRG) and Associate Professor of Computer Science at the Faculty of Computers and Information, Beni-Suef University. Dr. Elngar is a director of the Technological and Informatics Studies Center (TISC), Faculty of Computers and Information, Beni-Suef University. He is the managing editor of the *Journal of Cybersecurity and Information Management (JCIM)*. Dr. Elngar has more than 25 scientific research papers published in prestigious international journals and six books covering such diverse topics as data mining, intelligent systems, social networks, and the smart environment. Dr. Elngar is a collaborative researcher. He is a member of the Egyptian Mathematical Society (EMS) and International Rough Set Society (IRSS). His other research areas include the Internet of Things (IoT), Network Security, Intrusion Detection, Machine Learning, Data Mining, Artificial Intelligence. Big Data, Authentication, Cryptology, Healthcare Systems, and Automation Systems. He is an Editor and Reviewer of many international journals around the world. Dr. Elngar has won several awards including the "Young Researcher in Computer Science Engineering", from Global Outreach Education Summit and Awards 2019. Also, he was also awarded "Best Young Researcher Award (Male) (Below 40 years)" at the Global Education and Corporate Leadership Awards (GECL-2018). Dr. Elngar filed a patent on the "ElDahshan Authentication Protocol." Dr. Elngar has undertaken many activities in the community and the environment including organizing 12 workshops hosted by a large number of universities in almost all the governorates of Egypt.

N. Thillaiarasu currently works as an associate professor in the School of Computing and Information Technology, REVA University, Bangalore. He also served as the assistant professor in Galgotias University, Greater Noida from July 2019 to December 2020. He worked for seven years as an assistant professor in the Department of Computer Science and Engineering, SNS College of Engineering, Coimbatore. Dr. N. Thillaiarasu obtained his B.E. in Computer Science and Engineering from Selvam College of Technology in 2010 and received his M.E. in Software Engineering from Anna University Regional Center, Coimbatore, India in 2012. He received his Ph.D. from Anna University, Chennai in 2019, His areas of interest include Cloud Computing, Security, IoT, and Machine Learning.

T. Saravanan is currently an assistant professor in the Department of Computer Science & Engineering in GITAM University, Bengaluru Campus. He graduated with a B.E. in Computer Science and Engineering from Selvam College of Technology, Anna University, Chennai, India. He has a M.E. in Network Engineering from Anna University Regional Center, Coimbatore, India, and a Ph.D. in Information and Communication Engineering from Anna University, Chennai, India. He has 12 years of teaching experience. He has published many SCI papers, Scopus papers, book chapters, lecture notes, and Indian patents. His research interests include Wireless Sensor Networks, Artificial Intelligence, and Optimization algorithms. He has membership of MISTE, MCSI, IAENG, UACEE, and SDIWC.

Valentina Emilia Balas is currently full professor in the Department of Automatics and Applied Software at the Faculty of Engineering, "Aurel Vlaicu" University of Arad, Romania.

She holds a Ph.D. in Applied Electronics and Telecommunications from Polytechnic University of Timisoara, Romania. Dr. Balas is the author of more than 350 research papers in refereed journals and International Conferences. Her research interests are in Intelligent Systems, Fuzzy Control, Soft Computing, Smart Sensors, Information Fusion, Modeling, and Simulation.

She is the editor-in chief of the *International Journal of Advanced Intelligence Paradigms (IJAIP)* and the *International Journal of Computational Systems Engineering (IJCSysE)*. She is an Editorial Board member of several national and international journals and is an evaluator expert for national, international projects and Ph.D. Theses.

Dr. Balas is the director of the Intelligent Systems Research Center at Aurel Vlaicu University of Arad and the Director of the Department of International Relations, Programs and Projects in the same university.

She served as general chair of the International Workshop Soft Computing and Applications (SOFA) for eight editions 2005–2020 held in Romania and Hungary.

Dr. Balas has participated in many international conferences as an Organizer, honorary chair, session chair, and member in Steering, Advisory, or International Program Committees.

Now she is working in a national project with EU funding support: BioCell-NanoART = Novel Bio-inspired Cellular Nano-Architectures – For Digital Integrated Circuits, 3M Euro from National Authority for Scientific Research and Innovation.

She is a member of EUSFLAT and SIAM and a senior member of IEEE, a member of TC – Fuzzy Systems (IEEE CIS), chair of the TF 14 in TC – Emergent Technologies (IEEE CIS), and a member of TC – Soft Computing (IEEE SMCS).

Dr. Balas is a past vice-president (Awards) of IFSA International Fuzzy Systems Association Council (2013–2015), is a joint secretary of the Governing Council of Forum for Interdisciplinary Mathematics (FIM) (a multidisciplinary academic body in India), is a recipient of the "Tudor Tanasescu" Prize from the Romanian Academy for contributions in the field of soft computing methods (2019) and of the "Stefan Odobleja" Prize from the Romanian Academy of Scientists (2023).

Contributors

John Amose
KPR Institute of Engineering and
  Technology
Coimbatore, India

S. Boovaneswari
Manakula Vinayagar Institute of
  Technology
Puducherry, India

Ram Kumar C.
Sri Krishna College of Technology
  Coimbatore
Coimbatore, India

Vithya Ganesan
Koneru Lakshmiah Education
  Foundation
Vaddeswaram, India

G.K. Jagatheswari
PES University- EC Campus
Bengaluru, India

J. Jayapradha
Manakula Vinayagar Institute of
  Technology
Puducherry, India

Simarjeet Kaur
Chitkara University
Ambala, India

Chitra Kesavan
GITAM University, Bengaluru Campus
Bengaluru, India

Ajmeera Kiran
MLR Institute of Technology
Hyderabad, India

G. Ravi Kumar
CMR College of Engineering &
  Technology
Hyderabad, India

V. Sheeja Kumari
Saveetha School of Engineering,
  SIMATS University
Chennai, India

Dara Vijaya Lakshmi
REVA University
Bengaluru, India

G. Bindu Madhavi
Geethanjali College of Engineering and
  Technology
Hyderabad, India

R. Murugesan
Reva University
Bengaluru, India

Thillaiarasu N.
REVA University
Bengaluru, India

N. Palanivel
Manakula Vinayagar Institute of
  Technology
Puducherry, India

S. Ponmaniraj
Saveetha School of Engineering,
  SIMATS University
Chennai, India

Harshini R.
N.G.P. Institute of Technology
Coimbatore, India

M. Rajeshwari
Presidency University
Bengaluru, India

Viswanathan Ramasamy
Koneru Lakshmiah Education
  Foundation
Vaddeswaram, India

Y. Sowmya Reddy
CVR College of Engineering
Hyderabad, India

Vigneshwaran S.
Sri Ramakrishna Engineering College
Coimbatore, India

M. Saradha
REVA University
Bengaluru, India

C. Saravanan
Manakula Vinayagar Institute of
  Technology
Puducherry, India

G. Vennira Selvi
Presidency University
Bengaluru, India

A. Venkata Subramanian
GITAM University, Bengaluru Campus
Bengaluru, India

Sangeetha V.
Kalaignar Karunanidhi Institute of
  Technology
Coimbatore, India

Artificial Intelligence

1 Role of Artificial Intelligence in Telecommunication Systems
A Healthcare Perspective

*Sangeetha V., John Amose, Vigneshwaran S.,
Harshini R., Thillaiarasu N., and Ram Kumar C.*

1.1 INTRODUCTION

Artificial intelligence (AI) is intelligence exhibited by machines, as opposed to normal intelligence as shown by animals including people. The expression "artificial intelligence" describes machines that copy "mental" capacities that people associate with the human brain, for example, "learning" and "problem solving", AI applications encompass high-level web indexes (e.g., Google), suggestion frameworks (utilized by YouTube, Amazon, and Netflix), gathered human discourse (like Siri and Alexa), self-driving vehicles (e.g., Tesla), robotized direction, and competing at the most elevated level in essential game frameworks (like chess and Go) [1]. Artificial intelligence was established as a scholarly discipline in 1956, and in the years since has encountered a few surges in confidence, followed by dissatisfaction and the deficiency of financing (known as a "computer based intelligence winter"), followed by new methodologies, achievement and re-established financing. Artificial intelligence research has attempted and disposed of a wide range of approaches since its establishment, including recreating the mind, displaying human problem solving, formal rationale, huge information bases of information and mimicking animal behavior [2]. In the early years of the 21st century, advanced mathematical statistical artificial intelligence has dominated the field, and this strategy has demonstrated exceptional effectiveness, assisting with solving many testing problems through industry and the academic world.

The different sub-fields of AI research are based on specific objectives and the utilization of specific devices. The customary objectives of AI research incorporate thinking, information portrayal, arranging, learning, normal language handling, discernment, and the capacity to move and control objects. General intelligence (the capacity to take care of an inconsistent problem) is among the field's drawn out objectives. To tackle these problems, AI scientists have adjusted and coordinated a wide scope of

DOI: 10.1201/9781003388241-2

problem-solving procedures—including search and numerical advancement, formal rationale, artificial brain organizations, and strategies in view of insights, likelihood and financial matters. Simulated intelligence likewise draws upon software engineering, brain research, etymology, reasoning, and numerous different fields. The field was established with the understanding that human intelligence "can be so unequivocally depicted that a machine can be made to recreate it [3].

1.2 AI IN TELECOMMUNICATION

One more typical utilization of AI in media communications is building self-upgrading networks (SONs). Such organizations are consequently checked by AI calculations that distinguish and precisely foresee network inconsistencies. As demonstrated by IDC, 63.5% of telecom associations are actually executing AI to additionally foster their association establishment. PC-based intelligence in the telecom business uses advanced computations to look for plans inside the data, enabling Telcos to both recognize and expect network peculiarities. In light of including AI in telecom, CSPs can proactively fix issues before clients are oppositely impacted [4].

Numerous businesses value AI for its excellent capacity to examine large quantities of information. Given the telecommunications industry's continual access to vast amounts of data, it is not surprising that the combination of telecom and AI is more compatible than peanut butter and jam. How about we investigate the most widely recognized ways this innovation is utilized in media communications.

1.2.1 PREDICTIVE MAINTENANCE

As an organization develops and turns out to be more refined, keeping up with it turns out to be progressively troublesome. Fixing issues can be an expensive and tedious cycle. Also, it can prompt individual experiences and government intervention—something clients don't appreciate. Computer-based intelligence can have a major effect with prescient support. By tracking down designs in the verifiable information, AI and ML (Machine Learning) calculations can precisely identify and warn about conceivable equipment failures. This allows Telcos to be exceptionally proactive at maintaining their gear, fixing issues before they happen, and influencing the end-client [5].

Besides, these calculations can distinguish the cause of every failure, making it conceivable to battle the issue at its center. This occurred with one of the world's biggest suppliers of in-flight network and diversion, Gogo. They cooperated up with N-iX who worked on the nature of their in-flight web and made it conceivable to foresee hardware failures. Additionally, information science models utilized by the N-iX group distinguished the primary driver of malfunctioning antennas. Therefore, Gogo had the option to tackle the issue that was wasting money and causing individual moments.

1.2.2 NETWORK OPTIMIZATION

One more typical utilization of AI in media communications is building self-organizing networks (SONs). Such organizations are naturally checked by AI calculations that identify and precisely foresee network oddities. Moreover, they can proactively streamline and reconfigure the organization to guarantee that end-clients partake in continuous display and operation.

As organizations understand the benefit of involving AI in telecom network foundation, increasingly more will put resources into it. As per IDC, 63.5% of telecom organizations are effectively carrying out AI to further develop their organization framework [6].

1.2.3 VIRTUAL ASSISTANTS AND CHATBOTS

Conversational AI stages are one of the greatest forces to be reckoned with in the development of the AI in telecom market. These remote helpers, or chatbots, as they are additionally known, can computerize the treatment of client demands. Self-organizing networks having the ability to dynamically adapt is a desirable aspect because it allows for providing great customer service. This adaptive capability is highly advantageous for human-operated call centers. By scaling discussions to straightforward questions, chatbots can react to large numbers of client requests with amazing rate. This, in addition to the capacity to offer continuous support all day, every day, strongly emphasizes consumer loyalty. Vodafone saw an increment in consumer loyalty by 68% when they presented their chatbot TOBi.

As virtual assistants become more advanced and learn to handle more complex requests, the requirement for human administrators diminishes. This can assist companies with reducing their costs. Indeed, by 2024 the utilization of chatbots will prompt more than $8b in annual cost savings.

1.2.4 FRAUD DETECTION AND PREVENTION

The speech recognition and voice analytics market was valued at $20.98 billion in 2020 and is expected to grow at a compound annual growth rate (CAGR) of 15.4% between 2021 and 2028. Notwithstanding this, attacks on organizations actually cause more than $3.6B of damage yearly. Due to AI's scientific ability, many companies, including telecom, are considering using it to fight extortion. The most unmistakable benefit of AI-controlled extortion investigation is its capacity to forestall misrepresentation [7]. As soon as the system notices suspicious behavior, it stops the comparing client or management, stopping the lie from happening. Figure 1.1 shows that all of this is done in a way that makes the chances of not responding to an attack in time very small.

1.2.5 ROBOTIC PROCESS AUTOMATION (RPA)

RPA is a type of advanced change that depends on carrying out AI. Telcos can utilize RPA to mechanize information sections, request handling, charging, and other

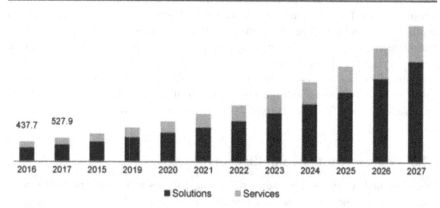

Asia Pacific telecom analytics market size, by component, 2016 - 2027 (USD Million)

437.7   527.9

2016 2017 2015 2019 2020 2021 2022 2023 2024 2025 2026 2027

■ Solutions   ■ Services

FIGURE 1.1   Global AI in telecommunication market, 2018-2023 (USD billion).

administrative center cycles that require loads of time and manual work. This opens up representatives' time, allowing them to zero in on more significant errands, and reduces the number of mistakes that physical work is inclined to make. Thus, the office runs smoother, workers are more useful, and clients appreciate error-free help [8].

Since there is so much to gain, it's not really a surprise that more than 53% of all businesses have already begun their journey in RPA. Additionally, this number is relied upon to develop to 72% in the following 2 years, while in 5 years robotic process automation (RPA) is expected to achieve widespread adoption across enterprises in various industries.

1.3   THE GROWING ROLE OF ARTIFICIAL INTELLIGENCE IN TELEHEALTH

Artificial Intelligence (AI) has turned into an ordinary reality as innovation advances. Medical services are one area that is rapidly changing on a major scale. From the issuance of electronic medical services cards to individual advice, telehealth is among the most up to date areas to utilize AI widely. AI is perhaps the main element forming telehealth in the United States today.

The utilization of AI in telehealth to permit specialists to make constant, information-driven decisions is a vital part in producing a superior patient experience and further developing well-being results as workers move up the consideration continuum, they offer more virtual consideration options (Figure 1.2). As indicated by MIT research, 75% of medical services offices that have already owned AI asserted it further developed their ability to cure disease, and four of every five said it assisted them with avoiding work burnout. With COVID-19 putting a strain on two areas (measure of clinical information and related patients, just as expanded specialist work), AI in telehealth is a solid choice for the eventual fate of clinical conveyance. This article talks about the way that AI is changing the telehealth scene.

FIGURE 1.2 Artificial intelligence in healthcare.

1.3.1 GIVING A MORE ACCURATE DIAGNOSIS

Telemedicine has made remote determination achievable. Experts can now look at, recognize, and treat sicknesses remotely. For instance, in patients with diabetic retinopathy, it has assisted with reducing visits. The L.A. Province Department of Health Services found that telehealth checking for diabetic retinopathy reduced patient visits by roughly 14,000 visits.

With the incorporation of AI to screening processes, the number of visits is expected to drop yet further. The calculation will possibly require retinal pictures assuming an AI framework is incorporated into the screening system. The AI framework can analyze the photos and contrast them with earlier examples to assess the seriousness of the illness with pinpoint exactness. The utilization of AI in the screening technique will save a lot of time and work for both the doctors and the patients.

Another organization is concentrating on pictures of patients to construct an AI framework to distinguish the presence of exceptional hereditary problems. Patients with exceptional hereditary disease presently normally need seven visits to the specialist's office before a legitimate diagnosis can be made. The number of visits might be brought down to zero utilizing AI and telemedicine. In essence an image of the patient's face is sent to the clinician, and the AI framework will assess it and precisely diagnose the condition. A conclusion could become a more time- and money-efficient way for both doctors and patients thanks to the ease of identifying evidence made possible by AI in telemedicine.

1.3.2 HOLDING DOCTORS BACK FROM BURNING OUT

During extended work days, specialists are inclined to become restless. It might cause disappointment and weakness, and affect work ability. Burnout is most frequently brought about by lengthy contact with patients or investing a lot of energy in innovative devices. Specialists engage with the Electronic Medical Record (EMR) instead than directly interacting with the patients. Telehealth has as of now become the standard for specialists in defeating burnout. At the point when AI and telehealth are joined, the benefits are multiplied. Artificial intelligence may help professionals save PC power. Simulated intelligence may likewise aid the discovery of signs that might propose burnout indications. It can even gauge the number of patients a specialist would have the option to visit prior to becoming depleted [9].

1.3.3 FURNISHING ELDERLY PATIENTS WITH BETTER MEDICAL CARE

Cell phones are frequently connected with telemedicine and there are numerous telemedicine applications created with the target of helping clients in dealing with their clinical issues, wellness objectives, specialist's visits, and protection guarantees more proficiently. Telehealth will finally join assistive robots. These robots will assume a huge part in conveying medical services to occupants, especially the elderly. Very smart robots can assist people with exercises like strolling and moving with regards to the home, conveying convenient drugs, and telling experts in case of a crisis.

The robots can move semi-independently and perform the responsibilities that have been relegated to them. The robots can understand the environmental elements, the patient's conduct and development designs, and the house inside settings on account of AI abilities. Thus, it might figure out how to help the patient and go about as their guardian [10]. The Japanese government has effectively started financing investigations into the production of AI robots to help the elderly. These robots can assist people with moving, washing, discarding their waste, and actually take a look at their well-being progressively. These robots can possibly bring down medical services conveyance costs while performing at a similar degree of proficiency as individuals. They might even help people improve their own happiness by thinking about the old.

1.3.4 PATIENT MONITORING CONVENIENCE

Computer-based intelligence has opened up a universe of opportunities for remotely checking a patient's well-being and mimicking up close and personal associations among specialists and patients. The utilization of man-made consciousness in telehealth settings is getting a foothold. Research on the adequacy of AI in remote diabetes care was performed by the Next DREAM Consortium Group. The fundamental outcome was that remote insulin changes utilizing the tested robotized AI framework might be just about as fruitful as learning measurement changes.

Doctors and experts might embrace the AI-based robotized administration for help. The Center for Telehealth Innovation at the University of San Francisco is exploring different avenues regarding AI that can decipher X-rays as an early notice sign for pneumothorax.

Information gathering and working together, patiently checking from afar, and coming to smart decisions and giving support are the most well-known AI use cases in telehealth. The capability of AI might be utilized to help specialists in diagnosing and treating patients, so as to lessen burnout and improve overall patient experience.

As the general health problem gets worse, medical services leaders are increasingly focusing on AI and telehealth as a way to stay competitive by making physician tasks easier and unlocking predictive potential through quiet information research. This has been seen as a decrease in the actual requirement for in-person discussions in 2023 because of the arrangement of lockdown measures. Artificial intelligence has altogether helped with the checking of customers' well-being in their homes.

1.3.5 MAKING HOSPITAL VISITS EASIER

In spite of the fact that telemedicine aims to eliminate emergency clinic visits, they are incidentally vital. In such circumstances, AI might help with reducing patient waiting times and guaranteeing that patients are viewed quickly. The AI innovation keeps staff individuals informed with regards to the inflow of patients visiting the medical clinic, high-need cases, the requirement for additional beds, and other patient-care-related data. The clinic's ability to acknowledge patients with muddled clinical issues has improved by 60% as a result of the program.

Rescue vehicle administrations have likewise further developed their productivity, with ambulances being sent an hour sooner. Because of AI and prescient investigation, patients in the emergency room are likewise delegated a bed up to 30% quicker in certain areas.

The significance of AI in telehealth will grow fundamentally as the area of telemedicine and telehealth receives more acknowledgment. The utilization of computerized reasoning in telehealth might be extremely advantageous to clinical specialists. It will assist with lessening costs, giving better medical services to individuals, and further developing the workplace by reducing patient wait times, suggesting the best therapy decisions, and making medical care available 24 hours per day, seven days per week.

1.4 DEVELOPING REQUIREMENT FOR VIRTUAL CONSIDERATION

As indicated by the American Telemedicine Association, around 200,000 individuals across the country get treatment in their homes by means of portable observing units—including telehealth units. Specialists say a maturing populace, expanding commonness of ongoing infections, the significant expense of medical services and innovative advances are powering development.

Harry Wang, a well-being research examiner with Parks Associates in Dallas, projects the more extensive home well-being observing business sector—for example, administrations and gear to follow on the off chance that an individual has fallen or isn't taking meds—will increase from $770 million in income in 2009 to $2.6 billion in 2023. The utilization of broadcast communications in medical services has expanded dramatically as of late and has opened doors for patients, medical care experts, and the executives of the well-being business overall. Telehealth, a term that alludes to the utilization of clinical data traded by means of electronic interchanges, extends patient access to better medical care and upgrades suppliers' jobs and capacities to supply better consideration [11].

4.1 TELEHEALTH AND TELEMEDICINE DEFINED

Among clinical innovation, telehealth incorporates electronic transmission of data that gives medical services experts more advanced working skills, such as continuing education, electronic medical record systems, and looking into working together.

The clinical side of telehealth is called telemedicine, characterized by the American Telemedicine Organization as "the utilization of clinical data traded starting with one webpage then onto the next by means of electronic interchanges to further develop patients' well-being status." Figure 1.3 shows the total telemedicine market in the United States from 2014 to 2025. Specific instances of telemedicine incorporate video discussions, remote patient information checking, nursing call focuses, and looking for or saving individual well-being data on the web.

1.4.2 BENEFITS FOR PATIENTS

Patients benefit greatly from telemedicine because it increases their access to care, educates them, facilitates research, and fosters innovation. Individuals in country regions and homebound patients can convey through phone, email or video meetings

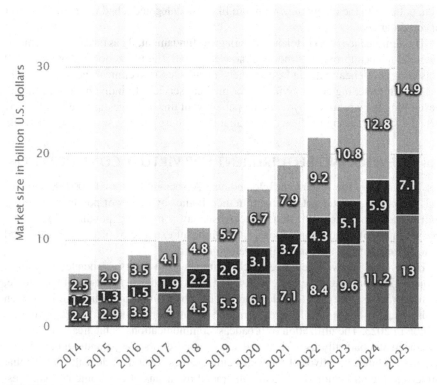

FIGURE 1.3 Total telemedicine market in the United States from 2014 to 2025, by product type (in billion USD).

with essential consideration specialists, medical attendants, and experts whose skill would not be accessible to them in any other case.

The Internet extends these capacities significantly further, managing the cost of the patient populace's access to lab results on the web, the capacity to explore their indications and conditions from home, access to clinical promotion gatherings, and the chance to profit from sites and stages intended to draw in and teach shoppers with well-being related interests [11].

1.4.3 Benefits for Providers

Suppliers benefit from health care services broadcast messages, as well as their association and the board of data, as well as electronic devices that help with clinical consideration. Electronic clinical records allow more straightforward recovery of patient data and are normally incorporated into charging frameworks and planning. People who work in health care can also continue their education and take classes online, and contact experts for sickness data outside his subject matter. In addition, he or she has more control over the board of health resolution and information on understanding progress. In a meeting with teachers who filled in as telemedicine caseworkers for adults with diabetes, the medical attendants and dietitians calling these patients announced satisfaction with both the practicality and viability of telemedicine. Benefits explicitly named were "more continuous contact with patients, more prominent unwinding and data because of the capacity to interface with the patients in their own homes, expanded capacity to come to the underserved, all the more opportune and precise clinical checking, and further developed administration of information" [12].

1.4.4 Strengthening and the e-Patient

This expanded access to data and the level of joint effort among patient and supplier has achieved a more understanding-driven medical care framework. Patients are looking into their conditions and drugs, assuming greater liability for their well-being and getting involved with online well-being related networks and informal organizations. This change has led to the expression "e-patient," portraying somebody who effectively searches out well-being data and correspondence on the web.

Benefits incorporate consistent encouragement and data sharing, access to doctor Q&A discussions, online records to catch progress and objectives and access to clinical preliminary data sets. A feeling of strengthening is the consistent idea going through these patient entries that associate patients to suppliers and to one another.

1.5 SIGNIFICANCE OF AI

- **Game Playing:** You can buy machines that can play ace level chess for a few hundred bucks. There is some AI in them, but they play well against people fundamentally through monster power estimation—looking at innumerable positions. To beat a title holder by monster power and acknowledged reliable heuristics requires having the choice to look at 200 million positions each second.

- **Talk Recognition:** During the 1990s, PC talk affirmation showed up at a helpful level for limited purposes. Thus United Airlines has traded its control center tree for flight information by a system using talk affirmation of flight numbers and city names. It is exceptionally worthwhile. On the other hand, while it is achievable to teach a couple of PCs using talk, most clients have gone back to the console the mouse.
- **Figuring out Natural Language:** Simple things like getting a list of words into a computer aren't enough. It is inadequate on the other hand to Parse sentences. The PC should be outfitted with an appreciation of the space the message is about, and this is eventually possible only for very limited regions [13].
- **PC Vision:** The world is made from three types of articles, but the commitments to the normal eye and PC TV cams are made in two steps. A couple of important tasks can work solely in two angles, yet full PC vision requires partial three-layered information that isn't just a lot of two-layered viewpoints. There are only a few limited ways to deal with three-layered material right now, and they're not as good as what people actually use.
- **Ace Systems:** A "data engineer" interviews experts in a particular region and endeavors to encapsulate their understanding in a PC program for doing some tasks. How well this functions depends upon whether the academic instruments required for the endeavor are in the present situation with AI. Right when this turned out not to be thusly, there were countless baffling results. One of the essential expert structures was MYCIN in 1974, which dissected bacterial pollutions of the blood and suggested prescriptions. It showed improvement over clinical students or practicing subject matter experts, gave its hindrances were taken note. Specifically, its cosmology included organisms, incidental effects, and meds and rejected patients, subject matter experts, clinical facilities, downfall, recovery, and events occurring on time. Its correspondences depended upon a single patient being considered [13]. Since the experts guided by the data engineers had some familiarity with patients, trained professionals, destruction, recovery, etc. clearly the data engineers took what the experts told them and put it all into a set template. The accommodation of current expert structures depends upon their clients having common sense.
- **Heuristic Classification:** With all the new information about AI, one of the most sensible ways to organize information as an expert is to use a few different sources to put some information in one of a nice set of groups. A model is telling us whether to accept a proposed Visa buy. Information about the Mastercard holder, his history of payments, what he is buying, and the store where he is buying it is available. For example, information about whether this store has had Visa fakes in the past is available.

1.6 ARTIFICIAL INTELLIGENCE IN TELEHEALTH

Another assessment of contemporary examples in telehealth proposed that two winning drivers for change were emerging: (1) high volume interest, on account of the growing difficulty to really help the patient, the clinician(s), and the connected

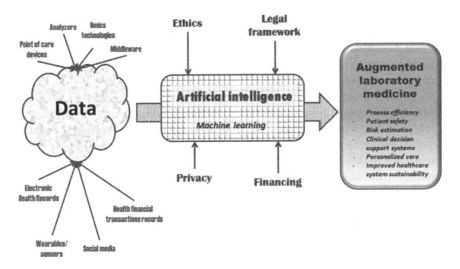

FIGURE 1.4 Utilization of artificial intelligence.

data, and (2) high criticality applications, where specific fitness is fundamental at the specific depiction of clinical premium in Figure 1.4.

Telehealth presents a couple of useful issues, for instance, when the media transmission interface misses the mark, or when, from a distance, the thought therapist isn't available. Reproduced intelligence could really clear up these situations by making it possible for human or virtual teams to work together, which would solve problems with scheduling and the availability of doctors (for instance, the time taken to appreciate the patient's anxiety or taking a bunch of experiences) [14].

Expanding progress is making the care course of action more unpredictable, past the degree of any single clinical provider. PC-based intelligence can maintain the progression of data on clinical cycles. Decisive reasoning and search, reasoning and acceptance, orchestrating, probabilistic reasoning and free bearing, learning, correspondence, knowledge, and mechanical innovation are all parts of the field of AI. They further proposed that PCs, in the space of artificial intelligence, could fill in as "sharp subject matter experts" and would reflect, if not match, individuals in points of view and thinking to the extent that PCs might mirror human mental lead and duplicate human mental execution. Taking into account this wide extent of expected results, Pacis et.al. summarized the possible impact of AI in telehealth around four rising examples considering unquestionable clinical consideration purposes: patient noticing, clinical benefits information advancement, adroit assistance and assurance, and information examination participation [15].

1.7 HOW IS ARTIFICIAL INTELLIGENCE UTILIZED IN MEDICATION?

Artificial Intelligence in medication is the utilization of AI models to look through clinical information and uncover bits of knowledge to assist with further developing well-being results and patient encounters. Because of late advances in software engineering and informatics, AI is rapidly turning into a basic component of current medical care. Simulated intelligence calculations and different applications controlled by AI are being utilized to help clinical experts in clinical settings and in continuous examination.

As of now, the most generally perceived positions for AI in clinical settings are clinical decision help and imaging assessment. Clinical choice helps apparatuses assist suppliers with settling on choices about medicines, drugs, psychological wellness, and other patient requirements by giving them fast access to data or exploration that is applicable to their patient. In clinical imaging, AI instruments are being utilized to examine CT filters, X-rays, MRIs, and different pictures for injuries or different discoveries that a human radiologist may miss.

The difficulties that the COVID-19 pandemic made for some well-being frameworks additionally drove numerous medical services associations all over the world to begin field-testing new AI advances, for example, calculations intended to assist with observing patients and AI-fueled apparatuses to screen COVID-19 patients.

The exploration and consequences of these tests are as yet being accumulated, and the general guidelines for the utilization AI in medication are as yet being characterized. However, doors for AI to help clinicians, scientists, and the patients they serve are consistently opening. Now, there is little uncertainty that AI will turn into a central component of the advanced well-being frameworks that shape and back current medication.

1.7.1 AI FOR INFECTION DETECTION AND DIAGNOSIS

Simulated intelligence in infection location and diagnosis. Unlike people, AI never needs to rest. AI models could be utilized to notice the essential indications of patients, providing fundamental attention and prepared medical professionals in the event that certain risk factors escalate. While clinical gadgets like heart screens can follow essential indicators, AI can gather the information from those gadgets and search for more perplexing conditions, like sepsis [16]. One IBM customer has fostered a prescient AI model for premature infants that is 75% precise in recognizing extreme sepsis.

1.7.2 CUSTOMIZED SICKNESS THERAPY

Precision in medicine could become more straightforward to help with virtual AI help. Since AI models can learn and hold states, AI can possibly give tweaked ongoing suggestions to patients nonstop. Rather than rehashing data with an individual each time, a medical services framework could offer patients nonstop access to an AI-fueled menial helper that could address questions in view of the patient's clinical history, inclinations, and individual necessities.

1.7.3 ARTIFICIAL INTELLIGENCE IN CLINICAL IMAGING

Artificial intelligence is now assuming a conspicuous part in clinical imaging. Research has shown that AI controlled by artificial neural organizations can be similarly viable as human radiologists at distinguishing indications of malignant breast cancer. As well as assisting clinicians with spotting early indications of infection, by identifying important aspects of a patient's experience and presenting relevant images, AI can help physicians manage the large amount of clinical images they must monitor..

1.7.4 CLINICAL PRELIMINARY EFFECTIVENESS

A lot of time is spent during clinical preliminaries allotting clinical codes to patient results and refreshing the important datasets. AI can assist with speeding up this interaction up by giving a faster and more insightful search for clinical codes. Two IBM Watson Health customers as of late observed that with AI, they could reduce their number of clinical codes searches by over 70%.

1.7.5 SPEED UP DRUG IMPROVEMENT

Drug discovery is one of the longest and most expensive parts of medication advancement. Computer-based intelligence could assist with diminishing the expenses of developing new meds in basically two ways: making better medication plans and tracking down promising new medication mixes. With AI, a significant number of the large information challenges confronting the existence sciences industry could be overcome.

1.8 ARTIFICIAL INTELLIGENCE IN MEDICAL DIAGNOSIS

Artificial Intelligence has become inseparable from help and proficiency in the medical local area. From an invention questioned as cases pushed it, AI has advanced to turn into the second pair of eyes that never need to rest. Clinical finding and clinical benefits of AI provides reliable support to exhausted clinical professionals and their workplaces, easing liability load and increasing expert viability. Clinical route, leaders, computerization, executive, and work processes benefit from AI in clinical determination. It can be used to analyze malignant growth, emergency medical imaging discoveries, banner intense anomalies, help radiologists focus on dangerous cases, analyze heart arrhythmias, predict stroke results, and manage chronic illnesses. Enhanced cognitive abilities is a rich area of data, computations, assessment, significant learning, brain associations, and pieces of information that are persistently creating and changing in accordance with the necessities of the clinical benefits industry and its patients. Over recent years, artificial intelligence in clinical finding has shown tremendous assurance in changing the standards of clinical thought while decreasing the ludicrous pressures felt by the clinical business.

1.9 ARTIFICIAL INTELLIGENCE IN MEDICINE

Artificial intelligence (AI) is being utilized in various areas of healthcare, including diagnostics, drug development, treatment personalization, and gene editing.

| Detecting lung cancer from CT Scans | Assess cardiac health from electrocardiograms | Classify skin lesions from images of the skin | Identify retinopathy from eye images |

FIGURE 1.5   Digitalization of diagnostic information.

1.9.1 DIAGNOSE DISEASES

Precisely diagnosing diseases requires significant clinical arrangement. Despite everything, diagnostics is routinely a difficult, tedious exchange of information . AI—particularly Deep Learning computations—have actually made tremendous advances in diagnoses, making diagnostics more affordable and more reachable [17].

1.9.1.1   How Machines Figure Out How to Diagnose

Machine learning computations can sort out some way to see plans similarly to how experts see them. A key difference is that computations need a lot of significant models—a huge number—to learn. Besides, these models ought to be helpfully digitized—machines can't figure out a more profound significance in understanding material. Along these lines, Machine learning is particularly helpful in locales where the indicative information an expert investigates is digitized.

Since there is a great deal of good data in these cases, computations are ending up being essentially something similar to the diagnostics as the trained professionals. What is important is the computation can arrive at conclusions in a fraction of a second and it may be repeated affordably in several locations Figure 1.5. Within a short period, individuals from any location can acquire a similar level of expertise in radiology diagnostics, and at a minimal expense [18].

1.9.2 DEVELOP DRUGS QUICKER

Creating drugs is an expensive collaboration. Drug advancement in wise cycles can be made more capable with Machine Learning. This might potentially shave off significant stretches of work and a huge number of hypotheses.

1.9.3 CUSTOMIZE TREATMENT

Patients react differently to drugs and treatments. Redid treatment may assemble patients' futures. Anyway, it's very challenging to perceive which factors ought to impact the choice of treatment.

1.9.4 IMPROVE QUALITY CHANGE

This method relies upon short helper RNAs (sgRNA) to target and change a specific region on the DNA. However, the associate RNA can fit different DNA regions - and that can provoke inadvertent delayed consequences (off-target influences).

1.10 PROS AND CONS OF ARTIFICIAL INTELLIGENCE IN HEALTHCARE

No matter what the business, AI has become common. As for prescription, AI assists prosperity specialists expedite tasks, improve utility, and design sophisticated methods. Tech giants are investing more in AI clinical research. AI collecting extends throughout the clinical benefits region, prompting questions about its benefits and cutoff thresholds [15–19].

1.11 CONCLUSION

Artificial intelligence can without a doubt carry new efficiencies and quality to medical services results in India. Be that as it may, holes and difficulties in the medical care area reflect well-established issues around lacking financing, powerless guidelines, inadequate medical services framework, and profoundly engrained socio-social practices. These can't be tended to by AI arrangements alone. The implementation of artificial intelligence is necessary in healthcare services, notably in the management of healthcare services, to go with clinical choices, particularly prescient investigation, in diagnosing and treating patients. Successful adoption of AI in the healthcare sector faces several critical challenges: managing user/provider adoption to new AI-driven systems and processes, ensuring AI technologies seamlessly integrate with existing healthcare infrastructure and workflows, proactively considering the patient perspective and experience, and optimally leveraging AI's capabilities across relevant use cases. Wellness, viability, security, data and assent, the choice to choose, "the option to attempt," expenditures, and access are among the moral issue records examined by AI clinical application.

REFERENCES

1. Barnes R, Zvarikova K. Artificial intelligence-enabled wearable medical devices, clinical and diagnostic decision support systems, and internet of things-based healthcare applications in COVID-19 prevention, screening, and treatment. *American Journal of Medical Research.* 2021;8(2):9–22.
2. Nanare IK, Panchwani BR, Gore N. Applications of Artificial Intelligence (AI) in telecommunication. *International Journal of Electronics, Communication and Soft Computing Science & Engineering (IJECSCSE).* 2018:12–7.
3. Liang G, Fan W, Luo H, Zhu X. The emerging roles of artificial intelligence in cancer drug development and precision therapy. *Biomedicine & Pharmacotherapy.* 2020 Aug 1;128:110255.
4. Ho D. Artificial intelligence in cancer therapy. *Science.* 2020 Feb 28;367(6481):982–3.
5. Bohr A, Memarzadeh K. The rise of artificial intelligence in healthcare applications. In *Artificial Intelligence in Healthcare.* 2020 Jan 1 (pp. 25–60). Academic Press.
6. Jeddi Z, Bohr A. Remote patient monitoring using artificial intelligence. In *Artificial Intelligence in Healthcare.* 2020 Jan 1 (pp. 203–234). Academic Press.
7. Conde JG, De S, Hall RW, Johansen E, Meglan D, Peng GC. Telehealth innovations in health education and training. *Telemedicine and e-Health.* 2010 Feb 1;16(1):103–6.
8. Harrer S, Shah P, Antony B, Hu J. Artificial intelligence for clinical trial design. *Trends in Pharmacological Sciences.* 2019 Aug 1;40(8):577–91.

9. Paul D, Sanap G, Shenoy S, Kalyane D, Kalia K, Tekade RK. Artificial intelligence in drug discovery and development. *Drug Discovery Today*. 2021 Jan;26(1):80.

10. Pesapane F, Codari M, Sardanelli F. Artificial intelligence in medical imaging: threat or opportunity? Radiologists again at the forefront of innovation in medicine. *European Radiology Experimental*. 2018 Dec;2(1):1.

11. Shantaram M. Impact of artificial intelligence in healthcare. *Biomedicine*. 2021 Oct 29;41(3):505–7.

12. Colombo S. Applications of artificial intelligence in drug delivery and pharmaceutical development. In *Artificial Intelligence in Healthcare*. 2020 Jan 1 (pp. 85–116). Academic Press.

13. Hassanzadeh P, Atyabi F, Dinarvand R. The significance of artificial intelligence in drug delivery system design. *Advanced Drug Delivery Reviews*. 2019 Nov 1;151:169–90.

14. Zhao M, Hoti K, Wang H, Raghu A, Katabi D. Assessment of medication self-administration using artificial intelligence. *Nature medicine*. 2021 Apr;27(4):727–35.

15. Miller DD, Brown EW. Artificial intelligence in medical practice: the question to the answer? *The American Journal of Medicine*. 2018 Feb 1;131(2):129–33.

16. Hamet P, Tremblay J. Artificial intelligence in medicine. *Metabolism*. 2017 Apr 1;69:S36–40.

17. Szolovits P, Patil RS, Schwartz WB. Artificial intelligence in medical diagnosis. *Annals of Internal Medicine*. 1988 Jan 1;108(1):80–7.

18. Ramesh AN, Kambhampati C, Monson JR, Drew PJ. Artificial intelligence in medicine. *Annals of the Royal College of Surgeons of England*. 2004 Sep;86(5):334.

19. Kuziemsky C, Maeder AJ, John O, Gogia SB, Basu A, Meher S, Ito M. Role of artificial intelligence within the telehealth domain. *Yearbook of Medical Informatics*. 2019 Aug;28(01):035–40.

2 An Intelligent System Utilizing Bipolar Fuzzy Logic for Ensuring Semantic Interoperability and Privacy Preservation in Healthcare Systems

M. Rajeshwari and R. Murugesan

2.1 INTRODUCTION

The expansion of Internet of Things (IoT) devices made specifically for the healthcare sector has opened up a wealth of opportunities. The enormous quantity of information produced by any of these connected devices seems to have the potential to fundamentally alter the healthcare sector. However, the personal data of healthcare information remains an important issue in the medical industry. Users are reluctant to disclose his\ her symptoms to the others, but many nations' laws require that patient information be maintained. The Health Insurance Portability and Accountability Act's (HIPAA) Privacy Rule lays out standards for people's rights to oversee and control how their patient data is used. Sensitive data regarding an individual, including details regarding their illness, their genealogy, their care plans, and their prescription drugs, to mention a few [1]. are referred to as PHI (Protected Health Information) by many nations, and every health center must keep it confidential.

Transmitting files from one place to another is essential whenever a patient needs to get an expert opinion from a specialist doctor or even if they are required to continue their therapies in another clinic. Given the varied characteristics of the IoT and hospital records, semantic interoperability in HIoT (Healthcare Internet of Things) increases. Conceptual interoperability refers to the capacity to transmit anything between systems without losing significance.

The requirement to effectively communicate medical data from one carer to another leads to the need for interoperability inside the healthcare sector. Different ehealth records use different medical words and terms, norms, as well as healthcare configurations. The capacity of the hospital setting to smoothly interact with some other hospital setting without sacrificing the real semantic information is known as

DOI: 10.1201/9781003388241-3

connectivity in the medical world. The terms IoHT (Internet of Healthcare Things) [2], IoMT (Internet of Medical Things) [3][4], HIoT[5], and Internet of Wellness Items [6] have all been used by various authors to refer to the Internet of Things in medicine. One of the biggest challenges confronting healthcare experts is the transfer of transparent electronic medical record.

It is important for most healthcare providers across India and other countries to adhere to open standards in the medical field, especially those related to health level seven (HL7) [7], open electronic health record EHR [8], and IHE-like standards in terms of document exchange formats. Interaction among healthcare professionals is made more challenging if two hospitals utilize various file styles. The files' utilization of different-coded medical phrases, including international classification of diseases for oncology (ICD-O) [9], international classification of diseases, tenth ICD-10 [10], systematized medical nomenclature for medicine–clinical terminology (SNOMED CT) [11], and logical observation identifiers names and codes (LIONC) [12], complicates medical technical problems.

Uncertain words can also be found in medical records. Pyrexia fever and hyperteat illness have remarkably similar concepts, and malignant tumor and tumor are generic terms that mean the exact thing. Through healthcare ontologies, the right and left eyes stand in for similar ideas. Medical documents are heterogeneous in terms of content, as well as their linguistic characteristics, such as malapropisms, homonyms, hyponyms, and hypernyms.

2.2 LITERATURE REVIEW

While one aspect of medicine strives to provide quick diagnoses and reasonably priced patient-centered treatment methods, the other side struggles with confidentiality and interoperability problems. The review of literature is divided into two sections because the study we are conducting addresses two issues.

Thillaiarasu and ChenthurPandian [13] provided examples of various uses for the Internet of Things in the healthcare industry. Mbengue et al. [5] assert that as digital healthcare terms have potential impact every day, their integration with different health documents will inevitably follow. A semantic interoperability challenge is presented by the integration of disparate medical history. A data structure was proposed by Jesus et al. [14] for storing and communicating data. SenML and OWL were synchronized in their proposed work. Cloud computing, fog computing, and edge computing were used to implement the system in a simulator. Semantic interoperability was not considered. Data from IoT devices was the only data they examined.

However, they did not propose a method for any other healthcare data, despite the fact that only a small percentage of the data was IoT data. As far as medical data privacy is concerned, no concerns have been raised. Thillaiarasu et al. proposed an OpenEHR extended ontology [15]. Furthermore, they included modules for authentication, publish–subscribe, probe, and translation in their framework called Medic (Medical Data Interoperability Through Collaboration of Healthcare Devices).

A framework based on ontologies for semantic interoperability was proposed by Ahamed et al. [5]. Medical records have linguistic issues that must be addressed in the proposed solution. Proge tool was used to develop a cardiovascular ontology so that semantic interoperability issues could be avoided among cardiovascular patients in different geographical locations.

In addition to six classes, the ontology contains five object properties, 23 data properties, and six stakeholders. The results were retrieved using SPARQL. Other diseases could not be included in the ontology since it only dealt with cardiovascular diseases. A semantic interoperability solution was provided by Sappagh et al. [16] for the development of a sound clinical decision support system.

A state-of-the-art approach for semantic interoperability depends on an approach based on ontologies [5], techniques that are based on rules [17],[18], agents [19],[20], and tabular documents [21]. Documents from a medical device, structured medical documents, or IoT documents are all the solutions available at present. A majority of linguistic problems with digital medical records must be addressed by these remedies.

The main techniques used throughout health coverage to safeguard patient confidentiality are access control [22],[23], anonymizing the data [24],[25], authentication[26],[27], and encryption [28],[29],[30]. Man-in-the-middle attacks frequently happen when using strong authentication. Different classes have implemented secure surface and transport layer security mechanisms to protect against such attacks. The utilization of authentication methods within the healthcare industry is prohibited by the following restrictions: (1) credentials should be kept by a third party; and (2) the system can only be accessed by registered users.

Data anonymization techniques have also been used to secure medicine information. The most used anonymity methods were K-anonymity and L-diversity. The anonymization process only partially protects the entire data set and anonymization is not appropriate. Authentication techniques are also used in the health sector to protect patient data. A well-known example of hash-based authentication is SHA-256. Offer the owner of the document must be available 24 hours a day, seven days a week.

Encryption key was also utilized to protect health information. To protect health information, an encryption key was used. The most widely used encryption algorithms were RSA, IDEA, and DES. Whenever the volume of data rises, the majority of medical systems upload it to the cloud.

To ensure the security of cloud storage, various encryption and authentication technologies, as well as attribute- and role-based access control schemes, have been created.

Despite the fact that different academics have proposed various techniques to ensure the privacy and security of healthcare data, the majority of existing solutions rely on a reliable cloud server. Some academics have proposed simple answers to semantic interoperability concerns in the medical industry, while others have suggested techniques to protect healthcare data. To the best of our knowledge, this is the first solution in the IoT for healthcare that combines privacy concerns with semantic interoperability concerns.

According to the latest reviewed literature, the following gaps had been determined.

1. The paradox in clinical files needs to be resolved due to the fact the prevailing systems primarily rely upon structured documents.
2. There are several relationships observed in the fitness statistics that need to

be handled, consisting of holonymy, meronymy, hypernymy, hyponymy, and polysemy.

3. The present-day cloud-based security alternatives rely upon a dependable record server.
   Therefore, the privacy of the covered fitness statistics (PHI) contained inside the medical documents could not be completely assured by the present strategies.

4. Currently there is no privacy keeping semantic interoperability solution for healthcare IoT in the literature.

2.3 PROPOSED SYSTEM

In this proposed system the three major building components are the document, encryption, and interoperability modules. There are three kinds of papers in the document module: structured, unstructured, and IoTMD. All are delivered to the interoperability module, which creates healthcare signature descriptive frameworks for every digital health record using the UMLS medical. Because the terminologies used in medical papers differ, the interoperability layer is in charge of resolving all semantic and lexical discrepancies inside the network. The interoperability module stores content vectors as reversed indexes.

Data privacy is ensured by the encryption module. The encryption module encrypts sign description frameworks and document vectors. It is not secure to use strong cryptography to encrypt documents, but it is possible to use symmetric encryption to do so, as Burman [31] suggests. Using both symmetric and asymmetric encryption methods, we proposed an approach for protecting healthcare data. Singular value decomposition is used for anonymizing the input query.

The key is protected via multiplicative homomorphic encryption using the asymmetric encryption algorithm ElGamal. Healthcare sign description frameworks or Electronic Health Records (EHR) are encrypted using acute encephalitis syndrome (AES). Several important concepts adopted for implementing the proposed work are described in the following subsections.

2.3.1 BFS AND SIMILARITY MEASURE OF CHOQUET COSINE

The concept of bipolar fuzzy set (BFS) in fuzzy set theory is similar to the traditional one as mentioned in Figure 2.1, but it takes into account both the degree of positive membership and the degree of negative membership of an element in a set. The degree of positive membership ranges from 0 to 1, while the degree of negative membership ranges from -1 to 0. This feature provides a more detailed representation of uncertainty and vagueness in data. The Choquet Cosine Similarity Measure is a variation in fuzzy set theory that specifically addresses fuzzy sets. It is defined as follows: Let $P = p_1, p_2, ..., p_n$ be a finite set, and let α and β be two BFS in P. Furthermore, let σ be a fuzzy measure on P. A Choquet cosine similarity measure between α and β can be defined as:

$$M_{BFS}^{(c,\sigma)}(\alpha,\beta) = (c)\int_P f_{\alpha,\beta}d\sigma \qquad (2.1)$$

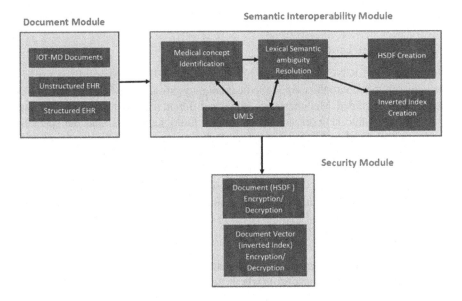

FIGURE 2.1 Architecture in general.

$$f_{\alpha,\beta}\left(p_i\right) = \frac{\mu_\alpha^P\left(p_i\right)\mu_\beta^P\left(p_i\right) + \mu_\alpha^N\left(p_i\right)\mu_\beta^N\left(p_i\right)}{\sqrt{\mu_\alpha^{P^2}\left(p_i\right) + \mu_\alpha^{N^2}\left(p_i\right)}\ \sqrt{\mu_\beta^{P^2}\left(p_i\right) + \mu_\beta^{N^2}\left(p_i\right)}} \quad \text{for } i = 1,2,\ldots,n \quad (2.2)$$

2.3.2 Decomposition of Singular Values

This method uses eigenvectors to reduce the dimension of the features. An n-by-n matrix M can be decomposed into three matrices: M_1, M_2, and M_3^T. M_1 is the left singular matrix, and its columns are the eigenvectors of $M^T M$ or MM^T. The singular matrix M_3^T with order m * m.

2.3.3 Computable Encryption

Any encryption technique which follows the commutative property of set theory is called commutative encryption. If E_1 is an encryption function, D_1 is a decryption function, G_1 and G_2 are two keys, input message and double encrypted message are represented as Q_{Mess} and double $R_{Message}$, respectively, then equations 1 and 2 hold for commutative encryption.

$$E_1\left(E_1\left(Q_{Mess}, G_1\right)G_2\right) = double\_R_{Message} \quad (2.3)$$

$$F_{\cdot_2}\left(F_{\cdot_2}\left(oduble\_R_{Message}, G_1\right)G_2\right) = Q_{Mess} \tag{2.4}$$

2.3.4 MEDICAL LANGUAGE

SNOMED CT, LIONC, RXNORM, ICD-O, and ICD-10 are some of the medical terms incorporated into the unified medical language system (UMLS). Every 2 years, the source is upgraded. Metathesaurus contains different medical terms and associated codes; the semantic network contains different biological terms; and the specialist lexicon with lexical tools contains lexical tools. An analysis of semantic networks shows the relationship between medical terms. The specialist lexicon repository in UMLS also contains some linguistic tools. UMLS tables are listed in Table 2.1. UMLS uses the following identifiers: AUI (Atom Unique Identifier), CUI (Concept Unique Identifier), LUI (Term Unique Identifier), and SUI (String Unique Identifier).

2.3.5 INTEROPERABILITY SEMANTIC MODULE

Various types of medical records and extracted key terms are presented to the semantic interoperability module. To ascertain whether a particular word is a medical expression or not, we derive the Concept Unique Identifier (CUI) from the UMLS ontology. The sequence of document preparation is illustrated. The semantic interoperability layer performs the tasks of resolving lexical variations, equivalent expressions, meronymy-holonymy issues, and hypernymy-hyponymy issues. Initially, the chunked words are scrutinized in the UMLS database to obtain a concept unique identifier (CUI). The UMLS's expert lexicon tools come with an inbuilt API provided by the National Library of Medicine that eliminates lexical variations.

2.3.6 CONFLICT RESOLUTION

Where a term elicits multiple CUIs, it is considered polysemous. In the pre-processing phase, we thoroughly examine all the polysemous terms and record all the relevant CUIs. A list of ambiguous terms is provided in [1]. If two distinct terms yield the same CUI, they are regarded as synonyms. All synonymous terms are assigned a single CUI. In natural language processing, meronymy refers to the relationship between a part and a whole, while holonymy represents the entire association. For instance, a toe

TABLE 2.1
UMLS Table

Objective	UMLS Table's
Ideas	MRCONSO
Relations	MRREL
Class divisions	MRHIER
Definition	MRDEF

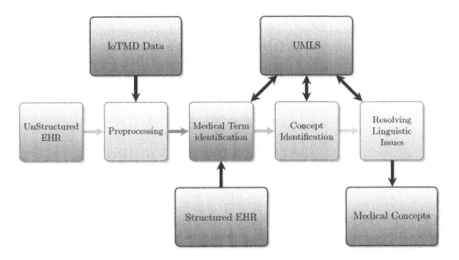

FIGURE 2.2 Recognition of medical concepts.

is a meronym of a leg, and a leg is a holonym of a toe. The MRREL table of UMLS is employed to retrieve the part-of relationships. If the MRREL field for a relationship indicates that a hypernymy-hyponymy relationship exists. The parent denotes the hypernymy, and the child denotes the hyponymy. The identification of medical concepts is demonstrated in Figure 2.2.

2.3.7 ASSOCIATION OF VAGUE SYNONYM SETS

In the field of healthcare, there exists an approximate association between words. For example, the term hyperthermia has a membership degree of 0.50 and 0.75 for the words temperature and high temperature, respectively. The MRSTY and MRCONSO tables' domains are utilized to establish a rule base for synonyms. The TUI and STN domains in the MRSTY table denote the unique identifier for the tree and the semantic type number, respectively. The formulation of the rule base for vague synonym sets is as follows.

$$\left(BFTUI\left(P_i\right) = T_x STN\left(p_i\right) = S.A.B.C.D TUI\left(p_j\right) = T_x STN\left(p_j\right)\right.$$
$$= S.A.B.C.D CUI\left(p_i\right) = CUI\left(p_i\right) then \mu^P S_{p_i}^{p_j}$$
$$= 1 and \mu^N S_{p_i}^{p_j} = -1 \tag{2.5}$$

$$\left(BFTUI\left(P_i\right) = T_x STN\left(p_i\right) = S.A.B.C.D TUI\left(p_j\right) = T_x STN\left(p_j\right)\right.$$
$$= S.A.B.C.D CUI\left(p_i\right) = CUI\left(p_i\right) then \mu^P S_{p_i}^{p_j}$$
$$= 0.74 and \mu^N S_{p_i}^{p_j} = -0.2 \tag{2.6}$$

$$\left(BFTUI\left(P_i\right) = T_x STN\left(p_i\right) = S.A.B.C.D TUI\left(p_j\right) = T_x STN\left(p_j\right)\right.$$
$$= S.A.B.C.D CUI\left(p_i\right) = CUI\left(p_i\right) then \mu^P S_{p_i}^{p_j}$$
$$= 0.5 and \mu^N S_{p_i}^{p_j} = -0.3 \tag{2.7}$$

$$\begin{aligned}(BFTUI\left(P_i\right) &= T_x STN\left(p_i\right) = S.A.B.C.D\,TUI\left(p_j\right) = T_x\,STN\left(p_j\right)\\ &= S.A.B.C.D\,CUI\left(p_i\right) = CUI\left(p_i\right) then\,\mu^P S_{p_i}^{P_j}\\ &= 0.25\,and\,\mu^N S_{p_i}^{P_j} = -0.5 \end{aligned} \qquad (2.8)$$

$$\begin{aligned}(BFTUI\left(P_i\right) &= T_x STN\left(p_i\right) = S.A.B.C.D\,TUI\left(p_j\right) = T_x\,STN\left(p_j\right)\\ &= S.A.B.C.D\,CUI\left(p_i\right) = CUI\left(p_i\right) then\,\mu^P S_{p_i}^{P_j}\\ &= 0.7\,and\,\mu^N S_{p_i}^{P_j} = -0.1 \end{aligned} \qquad (2.9)$$

2.3.8 FUZZY HYPERNYMY AND FUZZY HYPONYMY

Fuzzy hypernymy and fuzzy hyponymy refer to relationships between two medical concepts, where one concept is a partial subset or superset of the other. If two CUIs approximately capture this relationship, they are said to have a fuzzy hypernymy-fuzzy hyponymy connection, denoted as $(c_1, c_2, \mu_{hr}/\mu_{hp})$, where c_1 is an approximate subset of c_2 with a degree of μ_{hr}/μ_{hp}. The degree is represented as a value between 0 and 1 in the CM × CM domain. Fuzzy hyponymy captures partial subset relationships between two concepts, while fuzzy hypernymy captures partial superset interactions. These relationships are particularly useful in a system for recommending medications. For example, if Medicine B contains two different constituents with the same amount of medication as in Medicine A, and Medicine A comprises three distinct constituents, then Medicine B's fuzzy hyponymy is 0.70. To create a hypernymy rule base, the fields of MRSTY, MRREL, and MRCONSO tables are used. The TUI and STN fields of the MRSTY table represent three unique identifiers and semantic type numbers, respectively. The REL field represents the relationship in the MRREL table.

$$\begin{aligned}(BFTUI\left(P_i\right) &= T_x STN\left(p_i\right) = S.A.B.C.D\,TUI\left(p_j\right) = T_x\,STN\left(p_j\right)\\ &= S.A.B.C.D\,REL\left(p_i, p_j\right) = "bsf"\,then\,\mu^P \left(h_r\right)_{p_i}^{P_j}\\ &= 1\,and\,\mu^N \left(h_r\right)_{p_i}^{P_j} = 0 \end{aligned} \qquad (2.10)$$

$$\begin{aligned}(BFTUI\left(P_i\right) &= T_x STN\left(p_i\right) = S.A.B.C.D\,TUI\left(p_j\right) = T_x\,STN\left(p_j\right)\\ &= S.A.B.C.D\,REL\left(p_i, p_j\right) = "bsf"\,then\,\mu^P \left(h_r\right)_{p_i}^{P_j}\\ &= 0.75\,and\,\mu^N \left(h_r\right)_{p_i}^{P_j} = -0.25 \end{aligned} \qquad (2.11)$$

$$\begin{aligned}(BFTUI\left(P_i\right) &= T_x STN\left(p_i\right) = S.A.B.C.D\,TUI\left(p_j\right) = T_x\,STN\left(p_j\right)\\ &= S.A.B.C.D\,REL\left(p_i, p_j\right) = "bsf"\,then\,\mu^P \left(h_r\right)_{p_i}^{P_j}\\ &= 0.5\,and\,\mu^N \left(h_r\right)_{p_i}^{P_j} = -0.35 \end{aligned} \qquad (2.12)$$

2.3.9 HSDF (HEALTHCARE SIGN DESCRIPTION FRAMEWORK) DEVELOPMENT

To ensure semantic interoperability among healthcare IoT devices, it is essential to develop healthcare sign description frameworks. Our previous study [1] outlines the technique employed to create HSDF. Whenever documents pertain to crucial aspects like medication, vital signs, or symptoms, the corresponding healthcare sign description frameworks are developed.

2.3.10 MODULE FOR GENERATING EHR VECTORS

Medical documents are represented in this module as N-dimensional vectors. A row containing a feature's value at a specific time is stored in these vectors. TDM is computed using the indexer output. The TDM has medical terms as rows and medical records as columns. This matrix is then passed on to the module for creating the Latent Semantic Index. The semantic analysis module output is replaced by the Concept Document Matrix (CDM) with fewer dimensions in rows. Singular value decomposition is performed on the concept document matrix, resulting in three matrices. The first two matrices are encrypted using the Key KSV D, obtained from the Key Manager. EHR vectors are created by Principal Component Analysis. The encrypted and suppressed matrices are stored in Explorer. The process for creating EHR vectors is illustrated in Figure 2.3.

2.3.11 THE ENCRYPTION STAGE

The encryption process is composed of two stages: the encryption of the record, and the encryption of the record index. AES-128 key is employed to encrypt the records. ElGamal Cryptosystem is utilized to encrypt the key. The private hospitals

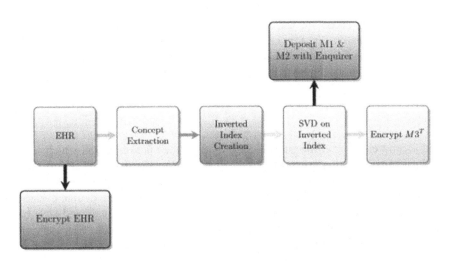

FIGURE 2.3 Creating EHR vectors.

determine a random key for the encryption of the EHRm,n record. The key used for the encryption of EHRm,n is encrypted using Keycommon, which is obtained from the key manager. Finally, each EHRm,n is saved in a trinity format (encrypted EHR, encrypted key, and EHR identifier), as presented in equation 2.13, with the storage manager.

$$E_{Doc}\left(EHR_{m,n}, Key_{doc}\right), EEI\left(Key_{Doc}, Key_{Common}\right), id\left(EHR_{m,n}\right) \qquad (2.13)$$

The Inverted Index of the Document is secured using AES-128 encryption. The key utilized to encrypt the index of the document is additionally secured through the application of ElGamal cryptosystem. The encrypted index is kept in Explorer as a pair of twins, represented by equation 2.14.

$$E_{id}\left[id\left(EHR_{m,n}\right), key_{id}\right], E_{EI}\left[Key_{id}, Key_{index}\right] \qquad (2.14)$$

2.3.12 PHASE OF DECRYPTION

Where a hospital j's client wants to access a document that belongs to another hospital i, hospital j creates a request based on the patient p's available data. To create the request, hospital j acquires M_2 and M_1 encrypted matrices from the inquirer and generates by decrypting the index key from the key manager; the concealed requests are then decrypted using equation 20. As soon as the Explorer receives this request, a double encrypting process is carried out by the hospital, and the encrypted key is forwarded to the key manager. A decrypted key is sent back by the Key Manager to the customer hospital j. The relevant indices can now be selected by hospital j after decryption. Figure 2.4 illustrates the process of retrieving the document identifiers.

The client medical facility transmits a group of EHR designators to the primary administrator for encoding. The encoded EHR designators are returned to the client. The client medical facility forwards the encoded designators to the repository administrator. The repository administrator computes the likeness and transmits a group of encoded records to the client in the layout indicated in formula 2.15.

$$E_{Doc}\left(EHR_{m,n}, Key_{doc}\right), E_{EI}\left[Key_{doc}, Key_{common}\right], id\left(EHR_{m,n}\right) \qquad (2.15)$$

The second element of the trinity is acquired by the client hospital, which is then subjected to encryption using their own $E_{EI}[Key_{doc}, Key_{common}]$ and transformed into a double encrypted key, as depicted in Equation 2.24. These keys are then transmitted to the key manager. Upon receipt, the key manager decrypts the double encrypted keys using the Key_{common} and forwards the resulting customer encrypted keys back to the customer. Finally, the customer hospital is able to retrieve the necessary plain documents, as illustrated in Figure 2.5.

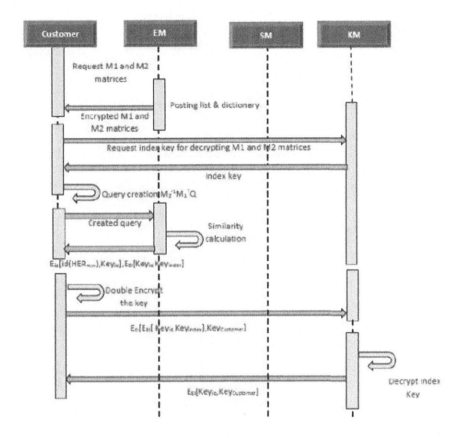

FIGURE 2.4 Retrieving document identifiers.

2.4 EXECUTION AND DEBATE

The setup is executed on Ubuntu 20.04.2 with a memory capacity of 8 GB and a hard disk space of 1 TB. The UMLS version 2019AA is installed on the Ubuntu machine and all RRF files are uploaded into MYSQL. The AES-128 algorithm is implemented using the Python pycrypto module in counter (CTR) mode. The Crypto public key package is utilized to execute ElGamal and generate random keys. The calculation formula presented in equation 15 is employed to determine the similarity between the encrypted documents and the query vector.

We investigated two evaluation aspects, one for evaluating safety and the other for evaluating meaningful compatibility. We scrutinized the security section by assessing computation time and anonymization. The efficiency of the retrieval system is utilized for evaluating meaningful compatibility. The proposed technique's swiftness is assessed by computation time. Anonymization techniques are employed for evaluating privacy. To evaluate meaningful compatibility, precision, recall, and F-measure are utilized. The following paragraph shows the CUI outcome obtained from UMLS.

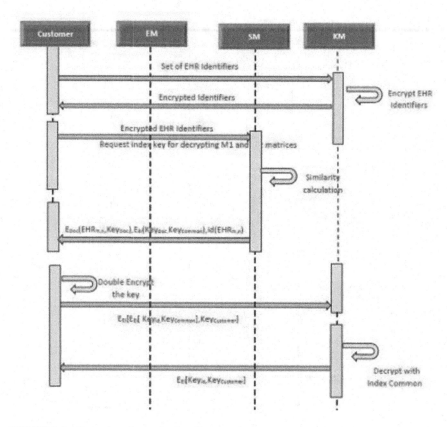

FIGURE 2.5   Procedure for retrieving documents.

When we input the term "temperature" into UMLS to search for the corresponding CUI, we get 244 results in the 2019AB release. Some of the CUIs that represent the term "temperature" include C0039476, C0005903, C4282227, and C0439237, which respectively correspond to "temperature," "body temperature," "temperature-degree," and "temperature-Celsius." Similarly, the CUIs for "pulse rate" and "normal pulse rate" are C0232145 and C0277900, respectively. In medical documents, there may be instances where monitoring data from IoT devices are abbreviated. For instance, the common name for medication "6 M.P" is not listed in UMLS. Instead, C0987634 and C0689485 correspond to the strings "Mercaptopurine 50 mg" and "Mercaptopurine 50 mg.oral," respectively.

Paracetamol and acetaminophen are interchangeable words as they both refer to the same CUI (C0793514). Certain medical terms have varying semantic categories. For instance, the term Hb has 927 distinct CUIs associated with it. Some of these CUIs pertain to the disease category while others relate to the enzyme category. When it comes to the term headache, the phrase intermittent headache is linked with C1168188 to represent the sign or symptom category. On the other hand, CUI C0752147 is associated with the phrase Chronic daily headache, which falls under the

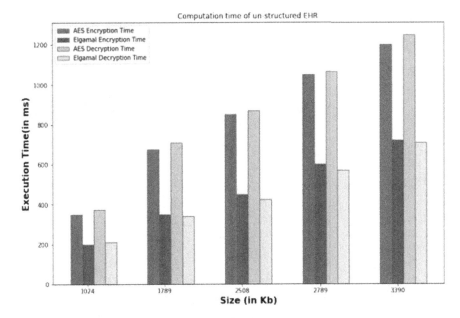

FIGURE 2.6   Documents unstructured execution time.

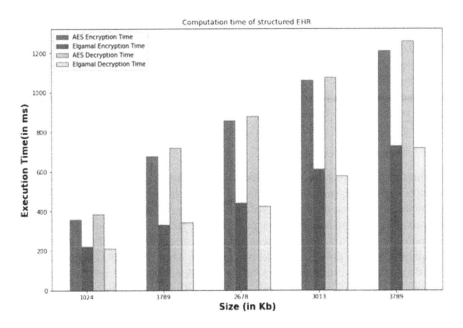

FIGURE 2.7   Structured execution time documents.

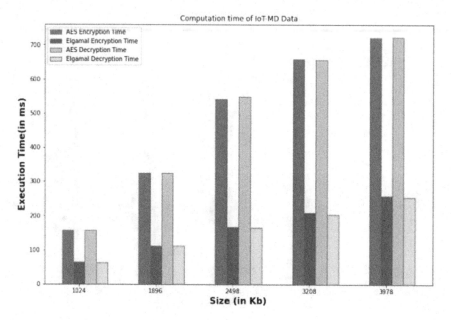

FIGURE 2.8   IoTMD execution time documentation.

disease or syndrome category. Documents unstructured execution time, Structured execution time documents and IoTMD execution time documentation are figured in Figure 2.6, Figure 2.7 and Figure 2.8. The representative outputs of UMLS are:

Thyroid 5Disease C0040125 is-a condition of Hypothyroidism C0020550. Pain in Right Leg C0564823 is a part of Pain in leg C0857031. Acetaminophen C0793514 is a synonym of Paracetamol C0793514. Hemoglobin C3841467 has multiple meanings and can refer to CO76175 Hemoglobin.

2.4.1   Time for Calculation

The time for calculation is the total duration of the formulation of the inquiry vector, generation of keys, encryption, and decryption. The formulation time for the inquiry vector is computed by adding up the time spent on locating the matrix's transpose, inverse, and multiplication. Index and document encryption times are combined to determine encryption time. The time taken to create the index and encrypt it is referred to as index encryption time. The time taken to decrypt the index involves double encryption time, index decryption time, decryption time at the Key Manager, and decryption time at the customer hospital. Document vectors require double encryption, key manager decryption, and customer hospital decryption time in order to be decrypted.

2.4.1.1   Time for Encryption

Time for encryption refers to the duration required for encrypting the records, encompassing the total time for encrypting electronic health records, encryption of indexes, and the time taken for formulating queries.

2.4.1.2 Time for Decryption

Time for decryption refers to the duration required to decrypt the initial records. This includes the total time for dual encryption of EHR indices, the total time for dual encryption of EHR vectors, decryption at the key manager's end, and decryption at the client's end.

2.4.1.3 Time of Communication

The time of communication refers to the duration taken for conveying the message. It encompasses the duration of data sensing, requesting of keyElgamal, time taken for conveying the Inverted Index to be saved in the Explorer, time spent for conveying the encrypted documents to be saved in the storage manager, time required for transmitting the encrypted vectors to the hospital making the request, time taken for conveying the double locked vectors to the key-manager from the hospital of the customer, and time taken for conveying the documents encrypted by the customer back to the customer.

2.4.1.4 Timeframe for Completion

The timeframe for completion is the duration between the submission of the request and the receipt of the unaltered EHR. It encompasses the time required for formulating the request, searching through the Explorer, forwarding locked documents to the customer, double-locking document vectors, transferring them to the Key Manager, converting double-locked documents into customer-locked documents, decrypting customer-locked documents, and requesting EHR vectors from the Explorer. The total processing time includes both communication time and the timeframe for completion.

2.4.2 TIME FOR UPDATING THE INVERTED INDEX

Figure 2.9 illustrates the duration required for updating the inverted index. To conduct the experiment, we utilized three files of equivalent size containing structured, unstructured, and IoTMD documents. The duration for inserting a single concept is insignificant for all three document types. However, removing a concept and inserting both the file and concept into the inverted index is considerably time-consuming, with no significant difference between the two. The overall time required for updating the index is higher for unstructured documents when compared to the other two types.

2.4.2.1 Linguistic Processing's Influence

Figures 2.10, 2.11, and 2.12 depict the precision graph, demonstrating the impact of linguistic processing. We assessed four retrieval systems for each category of document, namely, the standard text retrieval system, the standard text retrieval system with meronymy-holonymy resolution, the standard text retrieval system with meronymy-holonymy and hypernymy-hyponymy resolution, and the standard text retrieval system with meronymy-holonymy, hypernymy-hyponymy, and polysemy resolution. The fourth retrieval system (text retrieval system with meronymy-holonymy, hypernymy-hyponymy, and polysemy resolution) exhibited high accuracy

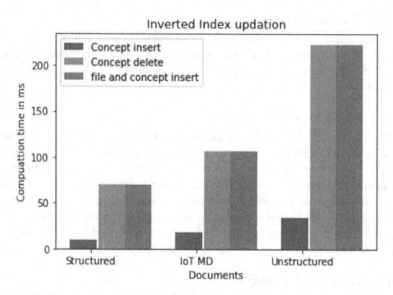

FIGURE 2.9 Time to update the inverted index.

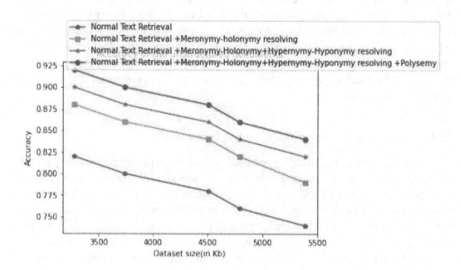

FIGURE 2.10 Reliability unstructured documents.

for all three types of documents. Eliminating the linguistic obstacles from medical documents enhances retrieval precision.

2.4.2.2 Comparison of Index Structures

Suppose that p denotes the quantity of EHR records, q is the number of search terms, and r signifies the number of terms contained in the query. The index structure utilized

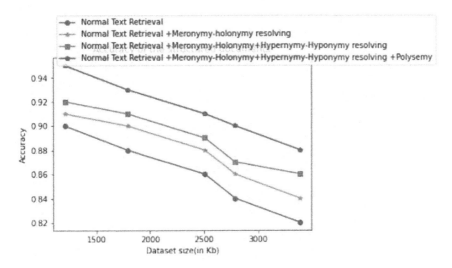

FIGURE 2.11   Accurately crafted documents.

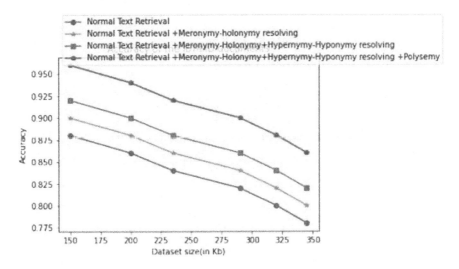

FIGURE 2.12   IoTMD documents accuracy.

by the [31] and [32] approaches is founded on a File-keyword structure, with an index magnitude of O(p).

In the [33], [34], and [31] methods all utilize a keyword file index structure with different index sizes, namely O(p.q), O(q2), and O(p.q), respectively. While the search complexity of the VPKE method is O(p), our index size is O(p*t) and the search complexity is O(p+t). It's important to note that the value of t, which represents the number of distinct CUIs in the system, is much smaller than q. This is because t is

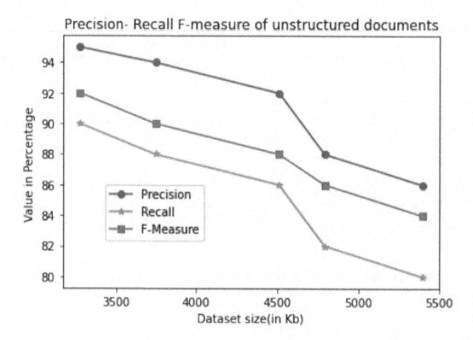

FIGURE 2.13   Unstructured documents with accurate recall.

determined after resolving issues such as duplicate entries, polysemy, meronymy-holonymy, hypernymy-hyponymy, and synonymous words.

$$\text{Exactitude} = |\text{ApplicableEHR} \cap |\text{FetchedEHR} / |\text{FetchedEHR}| \qquad (2.16)$$

$$\text{Completeness} = |\text{ApplicableEHR} \cap |\text{FetchedEHR} / |\text{ApplicableEHR}| \quad (2.17)$$

$$\text{Correctness} = 2. \text{Exactitude.Completenes} / \text{Exactitude} + \text{Completeness} \quad (2.18)$$

Retriever performance evaluation involves precision and recall measurements. Accuracy of the proposed system is assessed by precision, whereas completeness is evaluated by recall. Quality of a retrieval system cannot be determined by precision or recall alone. Therefore, the system's performance is measured by the F1 score or accuracy, which is the harmonic mean of precision and recall.

The plot shown in Figure 2.13 displays the Precision-Recall-F measure of IoT-MD. Six IoT MD documents with sizes of 150kb, 200kb, 235kb, 290KB, 320KB, and 340KB are considered. Precision values range from 98% to 88%, recall values range from 95% to 85%, and F-measure values range from 96% to 86%. As the document size increases, the precision and recall values, along with the F measure, decrease.

As shown in Figure 2.15, organized documents are represented by Precision-Recall-F Measure plots. Five organized documents with sizes of 1204 kB, 1189 kB, 2508 kB, 2789 kB, and 3390 kB are considered. Precision values range from 98% to 90%, recall values range from 94% to 86%, and F-measure values range from 95%

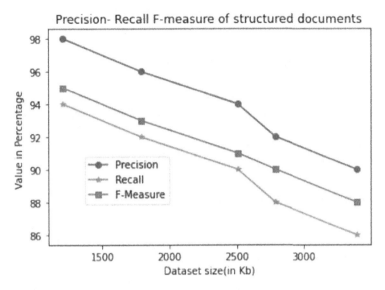

FIGURE 2.14 Documents with high precision recall.

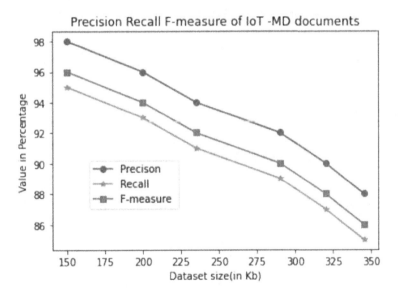

FIGURE 2.15 IoTMD documents with accurate recall.

to 88%. In addition to decreasing precision and recall values, the F-measure also decreases with increasing document size as shown in Figure 2.14. Figure 2.15 shows the Precision-Recall-F-measure plot of disorganized documents. Five disorganized EHRs with sizes of 3274 kB, 3743 kB, 4508 kB, 4791 kB, and 5394 kB are considered. Precision values range from 95% to 86%, recall values range from 90% to 80%, and

F-measure values range from 92% to 84%. As the document size increases, the precision and recall values, along with the F-measure, decrease.

2.4.2.3 Data Anonymization

The cloud-based Storage Manager is incapable of deciphering EHR documents that are in encrypted form. To ensure anonymity of both documents and document vectors at the cloud server level, commutative encryption is employed. The proprietor's chosen AES keys cannot be decoded by the Key Manager since these keys are doubly encrypted by the customer key. By reformatting the query input using equation $M_2^{-1}M_1^{T}Q$, the query input is also anonymized.

2.4.3 SECURITY ANALYSIS OF KEY PLAYERS

A randomly selected key is used to encrypt each EHR in the proposed system. All documents are encrypted in a way that only a legitimate client can decode them. Consumers can't decrypt other documents with the encryption key for one document, even if they are aware of that key. In the proposed system, the proprietor's role and that of the customer might differ. Neither a customer nor a proprietor can be the Storage Manager. It is not possible to obtain the encryption key since the Key Manager and Storage Manager cannot interact directly. While the Index Manager can interpret plain posting lists and dictionaries, it cannot receive plain or encrypted documents from storage managers or owners. The Explorer cannot determine the user's intentions as the customer enters the concealed query. Similarities between compressed queries and encrypted document vectors are calculated by the Explorer.

2.4.4 CONTRIBUTIONS

Our contributions involve constructing a model for semantic interoperability using the UMLS medical taxonomy. Subsequently, we generate inverted indices from this semantically transformed model. Finally, we secure EHR using a combination of AES and ElGamal encryption methods, ensuring that medical records are protected. Our study benefits from a hybrid encryption approach that incorporates both symmetric and asymmetric encryption techniques. Besides using bipolar fuzzy sets to increase accuracy, the proposed research model also uses Choquet cosine similarity measure to improve the efficiency. The UMLS ontology provides semantic interoperability while protecting the confidentiality of medical documents. The Index Manager creates the document index as an Inverted Index, making it easy to update. Because documents and indexes are maintained by two different parties, results of queries are anonymous. A distributed system maintains the privacy of medical records.

2.5 CONCLUSION

The expansion of IoT medical devices has resulted in challenges, such as compatibility and safety in the healthcare sector. To address both issues, a new framework has been suggested. The compatibility issue is resolved by developing a healthcare sign description framework using medical ontology UMLS. With the proposed system, all relevant keywords are given unique names based on CUI (Concept Unique Identifier), and concept-based retrieval is used to eliminate lexical and semantic ambiguities and decrease medical errors using bipolar fuzzy sets (BFS) and Choquet cosine similarity measures. AES and ElGamal encryption techniques are combined to protect medical data. Customers may choose which documents to decrypt and select based on the commutative property of the ElGamal encryption technique. Inverted indexes can be updated without having to be decrypted thanks to the homomorphic property of ElGamal's encryption method. Static interoperability in healthcare is our solution.

REFERENCES

[1] P. Sony, N. S, Semantic interoperability model in healthcare internet of things using healthcare sign description framework. *The International Arab Journal of Information Technology* 19(4), 589–596 (2022).

[2] //doi.org/https://doi.org/10.34028/iajit/19/4/3. https://iajit.org/portal/images/Year2022/No.4/21256.pdf

[3] S. Ketu, P. K. Mishra, Internet of healthcare things: A contemporary survey. *Journal of Network and Computer Applications* 192, 103, 179 (2021). https://doi.org/https://doi.org/10.1016/j.jnca.2021.103179. www.sciencedirect.com/science/article/pii/S1084804521001892

[4] A. Gatouillat, Y. Badr, B. Massot, E. Sejdić, Internet of medical things: A review of recent contributions dealing with cyber-physical systems in medicine. *IEEE Internet of Things Journal* 5(5), 3810–3822 (2018). https: //doi.org/10.1109/JIOT.2018.2849014

[5] S.M.K. Mbengue, O. Diallo, M.N. El Hadji, J.J.P.C. Rodrigues, A. Neto, J. Al-Muhtadi, in *2020 International Wireless Communications and Mobile Computing (IWCMC)* (2020), pp. 583–588. https://doi.org/10.1109/IWCMC48107.2020.9148130

[6] J. Ahamed, M.A. Chishti, Ontology based semantic interoperability approach in the internet of things for healthcare domain. *Journal of Discrete Mathematical Sciences and Cryptography* 24(6), 1727–1738 (2021).

[7] C.A. da Costa, C.F. Pasluosta, B. Eskofier, D.B. da Silva, R. da Rosa Righi, Internet of health things: Toward intelligent vital signs monitoring in hospital wards. *Artificial Intelligence in Medicine* 89, 61–69 (2018). https://doi.org/https://doi.org/10.1016/j.artmed.2018.05.005. www.sciencedirect.com/science/article/pii/ S0933365717301367

[8] R.H. Dolin, L. Alschuler, C. Beebe, P.V. Biron, S.L. Boyer, D. Essin, E. Kimber, T. Lincoln, J.E. Mattison, The hl7 clinical document architecture. *Journal of the American Medical Informatics Association* 8(6), 552–569 (2001).

[9] H. Leslie, International developments in open ehr archetypes and templates. *Health Information Management Journal* 37(1), 38–39 (2008).

[10] Fritz, April G., ed. *International classification of diseases for oncology: ICD-O.* World Health Organization, 2000.

[11] H. Quan, V. Sundararajan, P. Halfon, A. Fong, B. Burnand, J.C. Luthi, L.D. Saunders, C.A. Beck, T.E. Feasby, W.A. Ghali, Coding algorithms for defining comorbidities in icd-9-cm and icd-10 administrative data. *Medical care* pp. 1130–1139 (2005).

[12] K. Donnelly, et al., SNOMED-CT: The advanced terminology and coding system for health. *Studies in Health Technology and Informatics* 121, 279 (2006).

[13] N. Thillaiarasu, S. ChenthurPandian. "Enforcing security and privacy over multi-cloud framework using assessment techniques." *2016 10th International Conference on Intelligent Systems and Control (ISCO).* IEEE, 2016.

[14] P. Preethi, R. Asokan, N. Thillaiarasu, T. Saravanan, An effective digit recognition model using enhanced convolutional neural network based chaotic grey wolf optimization. *Journal of Intelligent & Fuzzy Systems* 41(2), 3727–3737 (2021).

[15] N. Thillaiarasu, S. C. Pandian, V. Vijayakumar, S. Prabaharan, L. Ravi, V. Subramaniyaswamy, Designing a trivial information relaying scheme for assuring safety in mobile cloud computing environment. *Wireless Networks,* 27, 5477–5490 (2021).

[16] Shenoy, Ashwin, N. Thillaiarasu. "A survey on different computer vision based human activity recognition for surveillance applications." *2022 6th International Conference on Computing Methodologies and Communication (ICCMC).* IEEE, 2022.

[17] E. Gothai, V. Muthukumaran, K. Valarmathi, V. E. Sathishkumar, N. Thillaiarasu, P. Karthikeyan, Map-reduce based distance weighted k-nearest neighbor machine learning algorithm for big data applications. *Scalable Computing: Practice and Experience* 23(4), 129–145 (2022). https://doi.org/10.3390/ s20226587

[18] A. Abbasi, B. Mohammadi, A clustering-based anonymization approach for privacy-preserving in the healthcare cloud. *Concurrency and Computation: Practice and Experience* 34 (2022). https://doi.org/10.1002/cpe. 6487

[19] H. Li, F. Guo, W. Zhang, J. Wang, J. Xing, (a,k)-anonymous scheme for privacy-preserving data collection in IoT-based healthcare services systems. *Journal of Medical Systems* 42, 56 (2018). https://doi.org/10.1007/ s10916-018-0896-7

[20] M. Sajjad, T.S. Malik, S. Khurram, A.A. Gardezi, F. Alassery, H. Hamam, O. Cheikhrouhou, M. Shafifiq, Efficient joint key authentication model in healthcare. *Computers, Materials Continua* 71, 2739–2753 (2022). https: //doi.org/10.32604/ cmc.2022.022706

[21] P. Huang, L. Guo, M. Li, Y. Fang, Practical privacy-preserving ECG-based authentication for IoT-based healthcare. *IEEE Internet of Things Journal* 6, 9200–9210 (2019). https://doi.org/10.1109/JIOT.2019.2929087

[22] D.N. Wu, Q.Q. Gan, X.M. Wang, Verififiable public key encryption with keyword search based on homomorphic encryption in multi-user setting. *IEEE Access* 6, 42, 445–453, (2018). https://doi.org/10.1109/ACCESS. 2018.2861424

[23] P.E. M. Vedaraj, A secure iot-cloud based healthcare system for disease classifification using neural network. *Computer Systems Science and Engineering* 41(1), 95–108 (2022). https://doi.org/10.32604/csse.2022.019976. www.techscience.com/csse/v41n1/44781

[24] Aroosa, Syed Sajid Ullah, Saddam Hussain, Roobaea Alroobaea, and Ihsan Ali. "Securing NDN-based Internet of health things through cost-effective signcryption scheme." *Wireless Communications and Mobile Computing* 2021 (2021): 1–13. https://doi.org/10.1155/2021/5569365

[25] S. Burman, A system using provably strong cryptography is probably not secure: Some practical attacks. *IETE Technical Review* 19, 161–168 (2002). https://doi.org/10.1080/ 02564602.2002.11417025

[26] M. Olgun, E. T'urkarslan, M. Unver, J. Ye, A cosine similarity measure based on the choquet integral for intuitionistic fuzzy sets and its applications to pattern recognition. *Informatica* 32(4), 849–864 (2021). https://doi.org/10.15388/21-INFOR460

[27] S. Purushothaman, N. Sureshkumar, Concept-based electronic health record retrieval system in healthcare IoT. *Advances in Intelligent Systems and Computing* 768 (2019). https: //doi.org/10.1007/978-981-13-0617-4 17

[28] S.J. Wu, R.D. Chiang, S.H. Chang, W.T. Chang, An interactive telecare system enhanced with IoT technology. *IEEE Pervasive Computing* 16(3), 62–69 (2017). https://doi.org/10.1109/MPRV.2017.2940967

[29] W.L.W.H.Y.W.B., Song, Inverted index based multi-keyword publickey searchable encryption with strong privacy guarantee. *2015 IEEE Conference on Computer Communications (INFOCOM)*, IEEE. pp. 2092–2100, 2015.

[30] R.Y.T.L.L...Z. R., Xue, Dynamic and efficient private keyword search over inverted index–based encrypted data. *ACM Transactions on Internet Technology (TOIT)* 16, 1–20 (2016).

[31] A. Gatouillat, Y. Badr, B. Massot, E. Sejdić, Internet of medical things: A review of recent contributions dealing with cyberphysical systems in medicine. *IEEE Internet Things* J 5(5):3810–3822 (2018).

[32] S.J. Nass , L.A. Levit, L.O. Gostin, (2009). *Beyond the HIPAA privacy rule: Enhancing privacy, improving health through research.* Washington (DC): National Academies Press.

[33] P. Huang, L. Guo, M. Li, Y. Fang, Practical privacy-preserving ecgbased authentication for IoT-based healthcare. *IEEE Internet Things* J 6:9200–9210 (2019).

[34] A. Jack, C. Percy, L. Sobin, S. Whelan, et al. (2000). *International classification of diseases for oncology.* World Health Organization.

3 Graph Optimizations in Neural Networks by ONNX Model

Viswanathan Ramasamy, Vithya Ganesan, and A. Venkata Subramanian

3.1 INTRODUCTION

3.1.1 GRAPH AND OPTIMIZATION

A graph is a collection of vertices (points) and edges (lines) where edges join the vertices. A graph can be directed and undirected graphs or weighted graphs or simple graphs. Graphs can be used to model a wide range of linkages and processes in physical, biological, sociological, and information systems and consist of many parameters. Optimization gives the best solution for the problem considering different parameters. Optimization strategies are a set of sophisticated instruments for effectively managing resources and maximizing the goal by optimization techniques such as graph rewriting, operation fusion, and scheduling.

In a graph-rewriting, a graph is replaced by another subgraph to derive a pattern graph for optimization. Operation fusion improves the speed by combining more operations and executing simultaneously without the requirement for a memory roundtrip. In the scheduling process of transferring a task graph to a destination platform, optimization improves the performance, ease of use, and cost savings. The goal of performance optimization is to reduce job execution time and speed up data processing. Ease-of-use optimizations are designed to make data processing more user-friendly. Cost optimization aims to maximize resource utilization and lower operational expenses. Graph optimization requires enhanced tools to increase optimization accuracy over a shorter amount of time. ONNX supports more computation and dynamic graph structures to improve the accuracy.

3.2 LITERATURE SURVEY

The Open Neural Network Exchange (ONNX) provides an exchange format for deep learning models and can be used to visualize or extract information from the underlying computational graph [1]. Deep Neural Network (DNN)-based applications are increasing in embedded edge devices [2]. To make different convolutional neural network models work with the ONNX format and give existing neural network models an

DOI: 10.1201/9781003388241-4

open standard interface, different convolutional neural network models and processes need to be used and compared with the framework [3].

In addition, ONNX Runtime includes several graph enhancements that help to increase model performance. Graph optimizations are modifications at the graph level, ranging from minor simplifications of graphs and nodes removal to more complex button merges and layout improvements [4].

Furthermore, the large amount of data with low locality causes an excessive increase in power consumption for data movement. Therefore, Processing-in-Memory (PIM), which places computing logic in/near memory, is becoming an attractive solution to solve the memory bottleneck of system performance [5].

All optimizations can be done online or offline. When starting an inference in offline mode, the ONNX Runtime serializes the resulting model to the hard disk followed by graphics optimization to reduce the warm-up time [6]. Machine learning mainly evaluates the cost function using labeled samples that are not easy to collect. Semi-supervised learning tries to find a better model based on samples without labels. Most semi-monitored methods are based on a graphical representation of samples and labels [7–17].

Graph-based semi-supervised learning focuses on least squares regression and cross-entropy classification. Optimization performance is measured by minimum validation cost and the maximum accuracy of the test to achieve the best cost.

3.3 METHODOLOGY

A novel optimization framework for graph-based semi-monitored learning is analyzed by the parameters such as CPU and GPU. The parameters such as inference time and latency time are compared by using PyTorch and ONNX model. Figure 3.1 shows the concept diagram of the ONNX model

3.3.1 GRAPH OPTIMIZATION IN ONNX

While loading a transformer model, ONNX Runtime performs most optimizations automatically. The following scenarios can benefit from this tool:

- Tf2onnx or keras2onnx export the model, and ONNX Runtime does not currently support graph optimization for them.
- Convert the model to float16 to improve performance on GPUs with Tensor Cores when utilizing mixed precision (like V100 or T4).

Concept model of ONNX

FIGURE 3.1 Concept model of ONNX.

- Due to form inference, the model has inputs with dynamic axes, which prevents several optimizations from being applied in ONNX Runtime.

3.4 RESULTS

Optimization parameters, such as CPU inference time between PyTorch and ONNX, are calculated as

CPU Inference time by PyTorch is shown in Figure 3.2.

ONNX Runtime of CPU inference time is shown in Figure 3.3.

```
[11] import time

    # Measure the latency. It is not accurate using Jupyter Notebook, it is recommended to use standalone python script.
    latency = []
    with torch.no_grad():
        for i in range(total_samples):
            data = dataset[i]
            inputs = {
                'input_ids':      data[0].to(device).reshape(1, max_seq_length),
                'attention_mask': data[1].to(device).reshape(1, max_seq_length),
                'token_type_ids': data[2].to(device).reshape(1, max_seq_length)
            }
            start = time.time()
            outputs = model(**inputs)
            latency.append(time.time() - start)
    print("PyTorch {} Inference time = {} ms".format(device.type, format(sum(latency) * 1000 / len(latency), '.2f')))

    PyTorch cpu Inference time = 510.93 ms
```

FIGURE 3.2   PyTorch CPU inference time.

```
[21] sess_options = onnxruntime.SessionOptions()


    sess_options.optimized_model_filepath = os.path.join(output_dir, "optimized_model_cpu.onnx")


    session = onnxruntime.InferenceSession(export_model_path, sess_options, providers=['CPUExecutionProvider'])

    latency = []
    for i in range(total_samples):
        data = dataset[i]
        ort_inputs = {
            'input_ids':   data[0].cpu().reshape(1, max_seq_length).numpy(),
            'input_mask':  data[1].cpu().reshape(1, max_seq_length).numpy(),
            'segment_ids': data[2].cpu().reshape(1, max_seq_length).numpy()
        }
        start = time.time()
        ort_outputs = session.run(None, ort_inputs)
        latency.append(time.time() - start)
    print("OnnxRuntime cpu Inference time = {} ms".format(format(sum(latency) * 1000 / len(latency), '.2f')))

    OnnxRuntime cpu Inference time = 389.64 ms
```

FIGURE 3.3   ONNX runtime of CPU inference time.

Figure 3.4 shows GPU Inference time by PyTorch.

Figure 3.5 shows GPU inference time.

Average prediction time per runtime is calculated between PyTorch and ONNX model as shown in Figure 3.6.

```
import torch

# Define your model and transfer it to GPU
model = YourModel().cuda()
model.eval()

# Prepare your input data and transfer it to GPU
input_data = torch.randn(batch_size, channels, height, width).cuda()

# Warm-up (Optional but recommended)
for _ in range(10):
    _ = model(input_data)

# Measure the time for inference
start_event = torch.cuda.Event(enable_timing=True)
end_event = torch.cuda.Event(enable_timing=True)

start_event.record()
with torch.no_grad():
    output = model(input_data)
end_event.record()

# Wait for the events to be recorded
torch.cuda.synchronize()

# Calculate and print elapsed time in milliseconds
elapsed_time_ms = start_event.elapsed_time(end_event)
print(f'Elapsed Time: {elapsed_time_ms:.3f} ms')
```

FIGURE 3.4   GPU Inference time by PyTorch.

```
import psutil
import onnxruntime
import numpy
import time
try:
    assert 'CUDAExecutionProvider' in onnxruntime.get_available_providers()
except AssertionError:
    exit(1)
device_name = 'gpu'

sess_options = onnxruntime.SessionOptions()

sess_options.optimized_model_filepath = os.path.join(output_dir, "optimized_model_{}.onnx".format(device_name))

sess_options.intra_op_num_threads=psutil.cpu_count(logical=True)

session = onnxruntime.InferenceSession(export_model_path, sess_options)

latency = []
for i in range(total_samples):
    data = dataset[i]
    ort_inputs = {
        'input_ids':  data[0].cpu().reshape(1, max_seq_length).numpy(),
        'input_mask': data[1].cpu().reshape(1, max_seq_length).numpy(),
        'segment_ids': data[2].cpu().reshape(1, max_seq_length).numpy()
    }
    start = time.time()
    ort_outputs = session.run(None, ort_inputs)
    latency.append(time.time() - start)

print("OnnxRuntime {} Inference time = {} ms".format(device_name, format(sum(latency) * 1000 / len(latency), '.2f')))

OnnxRuntime gpu Inference time = 1272.96 ms
```

FIGURE 3.5   GPU inference time by ONNX model.

```
[ ]  df.set_index('size')[['skl', 'ort']].plot(
         title="Average prediction time per runtime",
         logx=True, logy=True)
```

```
<matplotlib.axes._subplots.AxesSubplot at 0x7f60a7347950>
```

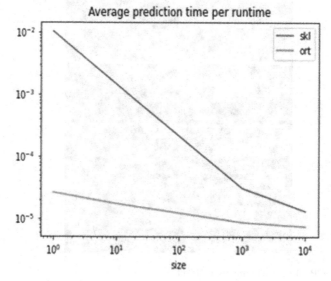

```
    print("Average prediction time per runtime",mt['skl'])
    print("Average prediction time per runtime",mt['ort'])
```

```
Average prediction time per runtime 1.2883096915999887e-05
Average prediction time per runtime 7.3269650740000996e-06
```

FIGURE 3.6   Average prediction time PyTorch and ONNX model.

3.4.1 IMAGE CLASSIFICATION

Image classification by the PyTorch method is shown in Figure 3.7 and its output is shown in Figure 3.8.

3.4.2 CONVERTING INTO ONNX

Based on the results time taken between PyTorch and ONNX is shown in Table 3.1.

```
outputs = ["TFLite_Detection_PostProcess", "TFLite_Detection_PostProcess:1",
           "TFLite_Detection_PostProcess:2", "TFLite_Detection_PostProcess:3"]

tflite_sess = interpreter_wrapper.Interpreter(model_path="ssdlite_mobiledet_edgetpu_320x320_coco_2020_05_19/fp32/model.tflite")
tflite_sess.allocate_tensors()
input2index = {i["name"]: i["index"] for i in tflite_sess.get_input_details()}
output2index = {i["name"]: i["index"] for i in tflite_sess.get_output_details()}

def infer_tflite(img):
    img = cv2.resize(img, (320, 320), interpolation=cv2.INTER_LINEAR)
    img = normalize(img)
    img = np.expand_dims(img, axis=0)
    tflite_sess.set_tensor(input2index['normalized_input_image_tensor'], tf.convert_to_tensor(img, np.float32))
    tflite_sess.invoke()
    result = [tflite_sess.get_tensor(v) for k, v in output2index.items()]
    box, label, score, num_detect = result
    label += 1
    return num_detect.astype('int32'), label.astype('int32'), score, box

one_image("000000088462.jpg", infer_tflite)
```

FIGURE 3.7　Image classification by PyTorch.

FIGURE 3.8　Output of image classification by PyTorch.

TABLE 3.1
Time Comparison between PyTorch and ONNX

Type of Model	Time for PyTorch	Time for After Conversion To Onnx
Bert Cpu	510 ms	389ms
Bert Gpu	1660ms	1272ms
Image Classification	16s	4s

FIGURE 3.9   Output of image classification by ONNX.

3.5   CONCLUSION

Traditional optimization is used to identify the best solution or unconstrained maxima and minima of continuous and differentiable functions. It uses mathematical methods and differential calculus to reduce or maximize the constrained optimization problems. Whereas ONNX Runtime includes considerable production-level optimization, testing, and other enhancements compared with existing algorithms and it is proven by the results as shown in Figure 3.9.

REFERENCES

[1]   Stier, J., and Granitzer, M. (2022). "Deepstruct – linking deep learning and graph theory." *Software Impacts*, 11, 100193. https://doi.org/10.1016/j.simpa.2021.100193

[2]   Lim, S. -H., Kang, S. -H., Ko, B. -H. Roh, J., Lim, C. and Cho, S. -Y. (2022). "Architecture exploration and customization tool of deep neural networks for edge devices." *2022 IEEE International Conference on Consumer Electronics (ICCE)*, pp. 1–2. doi: 10.1109/ICCE53296.2022.9730351

[3] Lim, Seung-Ho, et al. (2022). "An integrated analysis framework of convolutional neural network for embedded edge devices." *Electronics* 11.7, 1041.

[4] Junjie, B., Fang, L., and Ke, Z. et al. (2019). Onnx: Open neural network exchange GitHub Repository. https://github.com/onnx/onnx

[5] Kim, S. Y., Lee, J., Kim, C. H., Lee, W. J., and Kim, S. W. (2022). "Extending the ONNX runtime framework for the processing-in-memory execution." *2022 International Conference on Electronics, Information, and Communication (ICEIC)*, pp. 1–4. doi: 10.1109/ICEIC54506.2022.9748444

[6] Srilatha, Doddi, and Thillaiarasu, N. (2023). "Implementation of intrusion detection and prevention with deep learning in cloud computing." *Journal of Information Technology Management*, 15. Special Issue, 1–18.

[7] Shyamambika, N., and Thillaiarasu, N. (2016). "A survey on acquiring integrity of shared data with effective user termination in the cloud." *2016 10th International Conference on Intelligent Systems and Control (ISCO)*. IEEE.

[8] Thillaiarasu, N., and ChenthurPandian, S. (2016). "Enforcing security and privacy over multi-cloud framework using assessment techniques." *2016 10th International Conference on Intelligent Systems and Control (ISCO)*. IEEE.

[9] Preethi, P., Asokan, R., Thillaiarasu, N., and Saravanan, T. (2021). An effective digit recognition model using enhanced convolutional neural network based chaotic grey wolf optimization. *Journal of Intelligent & Fuzzy Systems*, 41(2), 3727–3737.

[10] Thillaiarasu, N., Pandian, S. C., Vijayakumar, V., Prabaharan, S., Ravi, L., and Subramaniyaswamy, V. (2021). Designing a trivial information relaying scheme for assuring safety in mobile cloud computing environment. *Wireless Networks*, 27, 5477–5490.

[11] Shenoy, Ashwin, and Thillaiarasu, N. (2022). "A survey on different computer vision based human activity recognition for surveillance applications." *2022 6th International Conference on Computing Methodologies and Communication (ICCMC)*. IEEE.

[12] Gothai, E., Muthukumaran, V., Valarmathi, K., Sathishkumar, V. E., Thillaiarasu, N., and Karthikeyan, P. (2022). Map-reduce based distance weighted k-nearest neighbor machine learning algorithm for big data applications. *Scalable Computing: Practice and Experience*, 23(4), 129–145.

[13] Kaladevi, A. C., Saravanakumar, R., Veena, K., Muthukumaran, V., Thillaiarasu, N., and Kumar, S. S. (2022). Data analytics on eco-conditional factors affecting speech recognition rate of modern interaction systems. *Journal of Mobile Multimedia*, 18(4), 1153–1176.

[14] Srilatha, Doddi, and Thillaiarasu, N. (2022). "DDoSNet: A deep learning model for detecting network attacks in cloud computing." *2022 4th International Conference on Inventive Research in Computing Applications (ICIRCA)*. IEEE.

[15] Srilatha, D., and Thillaiarasu, N. (2022, September). OIDCBMS: A novel neural network based intrusion detection system to cloud computing based on new cube algorithm. In *2022 4th International Conference on Inventive Research in Computing Applications (ICIRCA)*, pp. 1651–1656. IEEE.

[16] Zhou, Yu, et al. (2022). "Graph neural networks: Taxonomy, advances, and trends." *ACM Transactions on Intelligent Systems and Technology (TIST)*, 13.1, 1-54.

[17] Tu, Enmei, Wang, Zihao, Yang, Jie, and Kasabov, Nikola. (2022). Deep semi-supervised learning via dynamic anchor graph embedding in latent space, *Neural Networks*, 146, 350–360. https://doi.org/10.1016/j.neunet.2021.11.026

4 Convolutional Neural Network Architecture for Accurate Plant Classification

N. Palanivel, J. Jayapradha, S. Boovaneswari, and C. Saravanan

4.1 INTRODUCTION

The agriculture sector is one of the most essential fields for the success of a nation's economic growth. There are a number of categories of plants across the planet: medicinal plants, flower plants, fruit-giving plants, seed plants, grass, vines, etc. The plants that can be considered the primary source of medicines are called medicinal plants. These plants can be used for varied purposes [1]. The identification of a plant heavily depends on deep research on the plant's botanical characteristics. Comprehensive botanical research can improve classification accuracy and can avoid ambiguities and errors in plant species classification [2-5]. The detection of plants can be performed based on the feature comparison to earlier gathered and categorized samples. All the samples are collected based on the sample properties and represent a certain category of plant species. However, the identification and classification of plant species is a complex and challenging process due to their structures, shapes, color, and lifestyles. Thus, a deep classification of these plants is required for plant analysis and understanding. Moreover, the process of plant detection is influential on botanical nomenclature and plant taxonomy [2]-[5]. Thus, a speedy and automated plant species detection and classification method are useful for identifying the class of a particular plant. Plant species detection is very useful for the management of plant phenotyping and plant agriculture [6-7].

However, there are several types of classification performed by various researchers and botanical experts such as phytochemical classification, plant genetic classification, plant serum classification, and plant cell classification. However, these types of classification are critical tasks for researchers or professionals due to operating complexities and poor practicality. The classical methods utilized for plant species detection and classification do not automate the classification process and provide limited efficiency results. Thus, efficiency improvements in terms of detection and classification performance require a wealthy taxonomic knowledge of plants and an adequate classification mechanism. Another problem area is the process of data collection,

DOI: 10.1201/9781003388241-5

as most of the classification methods rely upon public datasets and a manual data acquisition process is adopted in these large datasets. Thus, classification accuracy and classification objectivity can be affected due to the low labor efficacy, and hefty workload. However, one of the solutions to this problem is the faster development and advancement of computer image processing and pattern detection technologies. These technologies can provide massive benefits in identifying plant species quickly and precisely, even in the case of manual data acquisition. Thus, understanding plant species and the classification of obtained plant features are of great importance based on digital plant photographs. These digital plant images are massively useful in the case of plant disease recognition, plant leaf identification, plant diversity protection, and plant species detection. Thus, deep learning methods can be massively beneficial for the identification of plant class and plant species classification through digital plant images. The large success of deep learning methods in image processing applications is due to their biological neurons' functions and structure. Another advantage of deep learning methods is adequate data processing and substantial improvements in the recognition of features and patterns. Thus, some of the research works related to plant detection and classification are presented in the following.

Haque and Adolphs [8] presented a plant disease detection and classification methodology based on deep neural network architecture. In this architecture, data augmentation is performed with transfer learning for plant disease classification. Hyper-parameters are tuned to identify disease locality. Testing of the model is performed on the Corn leaf infection dataset. Data preprocessing is performed to enhance classification performance. Gajjar et al. [9] presented a plant identification method based on one of the Convolution Neural Network Architecture (CNN) models. This architecture mainly focused on reducing the misbalancing effect in a classification process. The model was tested on different datasets such as Swedish, Leaf Snap, Folio datasets, Middle European, Flavia, and Woody Plants 2014. Zhange et al. [10] adopted a transfer learning approach to detect plant diseases based on CNN-enabled Multi-Pathway Activation Function (MPAF). The varied diseases that can be identified using the given transfer learning approach are maculopathy, blight, and rust. Furthermore, the MPAF model is utilized to improve the performance and accuracy of the model. Srilatha et al. [11] performed the detection of plant class and plant leaf diseases using transfer learning and CNN architecture; 14 varied classes of plant species were utilized for training in plant classification. In addition, 38 varied categorical classes related to the disease were identified and the model provided better performance results than different classical mechanisms.

However, the above-mentioned literature is primarily focused on plant leaf disease detection and its classification. Until now, only a few research works have focused on plant species identification or plant class detection among a variety of species or classes available around the world. Therefore, planet species identification is performed in this research work using a proposed Convolution Neural Network Architecture (CNN)-based Plant Classification (CNN-PC) model. The proposed CNN-PC model accurately identifies which plant image belongs to which species among several available plant species. The deep learning model allows precise identification of plant features and optimization of training models using different layers and blocks of the proposed CNN-PC architecture. These layers can be sequential layers, pooling layers, flatten layers, drop-out layers, fully linked layers, etc. Every layer has been assigned some specific task to achieve accurate plant classification. The proposed CNN-PC model can be segregated into different phases for better analysis of plant images such as the data acquisition phase, pre-processing phase, essential feature selection, extraction phase, training phase, testing phase, and classification performance evaluation phase. The proposed CNN-PC model is tested on a large Vietnam plant dataset for evaluating testing and classification performance. This work is presented in the following style. Section 4.2 discusses the mathematical modeling of the proposed CNN-PC model for plant species identification. Section 4.3 discusses the simulation results and their comparison with traditional plant classification techniques, and section 4.4 concludes the present work.

4.2 MODELING OF PROPOSED CNN-PC MODEL

This section discusses a detailed methodology of the proposed CNN-PC model for the detection of plant species based on classification performance. The main focus of this research work is the classification of plant images and detection of plant species using the proposed CNN-PC model. Very few classification methods are focused on this purpose. The proposed CNN-PC model efficiently detects which plant image belongs to which class with high-performance efficiency. Figure 4.1 demonstrates

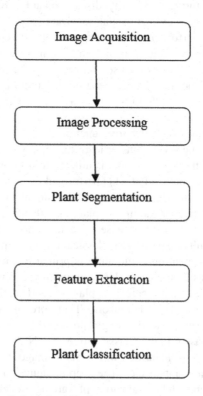

FIGURE 4.1 Flowchart of plant classification.

the basic working methodology of the proposed CNN-PC model which includes the presence of varied layers and blocks.

4.3 SYSTEM ARCHITECTURE

The model presented aims to accurately design a classification model based on multiple layers such as sequential layers, pooling layers, flatten layers, drop-out layers, and fully linked layers. Along with that convolutional and dense blocks are utilized to develop an efficient plant classification model shown in Figure 4.2. The optimization of hyper optimization training parameters is achieved using some optimization layers such as Adam and RMSProp optimizers. These optimizers can enhance the estimation ability and quality of training weights. So that, overall classification performance accuracy can be improved.

First of all, a large dataset is selected related to plant images to perform effective training. After the selection of data, training and testing images are segregated. Here, 100% of images are used for training, and 60% of images are utilized for testing. After the dataset fixing, input data is fed to the proposed CNN-PC model and pre-processing is performed to minimize errors and ambiguities in dataset images and analyze given input images. It also generates pre-trained information which contains discriminative information related to the image size, image transformation (vertical or horizontal) flipping, total number of training, validation, and testing images, scaling, number of classes, and number of epochs hyper-parameter tuning, etc. Based on this information and the performed analysis, convolution neural network architecture is designed, such as the number of layers to be used, kind of layers to be used, block size, layer dimensions, and model design. The layers and blocks that are used in the proposed CNN-PC model are sequential layers, pooling layers, flatten layers, drop-out layers,

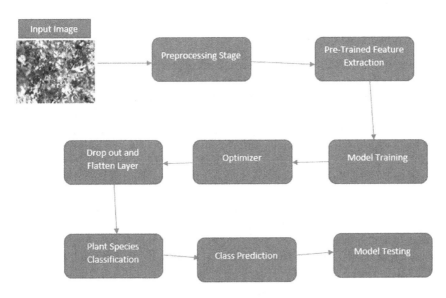

FIGURE 4.2 System architecture.

fully linked layers, and dense and convolutional blocks, respectively. After model design, training weights are generated and efficient model training is performed considering certain hyper-parameters. Hyper-parameter tuning is required to improve classification accuracy. Then, generated feature maps in training can be utilized to predict labels in accordance with the ground labels. Then, model testing is performed by comparing ground truth labels and predicted labels so that a confusion matrix is obtained. Based on confusion matrix parameters, performance metrics like classification accuracy, precision, recall, and F1-measure parameters are evaluated. Finally, the evaluated predicted labels from input images after training of the model are used for plant species detection among various plant species accurately. Furthermore, the proposed CNN-PC model minimizes computational complexity, generates dataset uniformity, provides image smoothening, obtains feature enhancement results, and performs accurate classification predictions.

The key focus of the proposed CNN model is a precise understanding of plant image data with the minimum computational resource utilization. The obtained pre-trained weights contain pixel information that is nearest to the central position of an image. An image is transformed multiple times by horizontal flipping, vertical flipping, shearing, and zooming, rotation by different angles in different epochs so that efficient plant classification can be performed. All the layers are grouped in three different sets. Here, set 1 represents a combination of different layers like a convolutional layer, batch normalization layer, and a ReLU activation function. Similarly, set 2 consists of two different types of pooling layers. And the last set contains a fully linked layer, a soft-max layer, and a classification layer. Moreover, the dependencies related to Tensor Flow and Keras are used for data processing and feeding data into the network. Furthermore, a detailed mathematical representation of some of the essential building blocks of the proposed CNN-PC model are convolutional layers, sequential layers, batch normalization layers, pooling layers, flatten layers, and fully linked layers.

4.4   CNN-PC ALGORITHM

Convolutional layers are one of the prime components of the proposed CNN-PC model and contain multiple layers which help to generate feature weights. These layers can perform multiple tasks such as detection of edges, removal of blurring effect, sharpening the given image, and region of interest identification. Here, fine-grained discriminative features help build an efficient trained classification model with higher accuracy. There can be different layers in the dense block and the size of hidden output arrays can vary from 4 to 32 neurons. Different sizes of convolutional layers are utilized and the ultimate aim is to reduce the dimensions of these layers so that minimum computational resources are utilized. Another advantage of using convolutional layers is their ability to perform merging operations on two different sets of information. Feature weights can be generated using convolutional filters on input data and those generated weights can be mapped into feature maps. Then, the generation of feature weights from these layers is given by the following equation:

$$K_C = \lambda \left(K_{t-1} * G_t + h_t \right) \qquad (4.1)$$

where the given input image is expressed by G_z and K_t is represented as feature weights. Then, λ is defined as the ReLU activation function and the bias coefficient is represented by h_t. Finally, the output feature map is expressed as K_t and represents a convolutional operator. Each of the many convolutional filters that make up convolutional holds its own unique set of information. All the information is summed together to generate fine-grained discriminative features from given input images. These features are constantly updated in the training so that plant image classes can be differentiated efficiently. Each image is transformed and flipped multiple times to get feature weights for each image and all the features weights obtained from each image are summed up to get final feature maps. Hyper-parameter tuning speeds up the training of the proposed CNN model. Computational cost can be minimized by reducing the size of filters. Then, the total loss that occurs in the feature extraction process is given by the equation:

$$G(t,d,y,b) = M^{-1}\left[G_d(t,d) + K_t G_{lo}(t,y,b)\right] \qquad (4.2)$$

where G_d is represented as a validation loss and K_t represents generated feature weights. Moreover, M represents the number of training iterations. Pixel localization loss is given by G_{lo}. Input images are fed to the convolutional layers and convolution filters or kernels are used to generate filter shape. Then, convolutional operations are performed by performing element-wise matrix multiplication for each image, and unique weights are generated for each image. All the generated weights are summed and a feature map is obtained.

The batch normalization layer is used for the normalization of contributions to a layer in each mini-batch to perform deep training of the proposed CNN model and reduces the total number of training iterations. Practical parametrizing of the proposed CNN model and output layer scaling is performed by normalizing activations of each input in each mini-batch. Normalization rescales data to zero mean and one standard deviation

$$c = \left[(\tau - t)(J^2 + \beta)^{\frac{1}{2}}\right].\Psi + \psi \qquad (4.3)$$

where standard deviation and mean are represented the current epoch. After every epoch, training hyper-parameters get updated. Here, β is a small constant to avoid zero division. ReLU activation function is employed for object identification and classification using the proposed CNN model. Then, the ReLU activation function is represented by the equation:

$$f(G_t) = \begin{cases} G_t, & \text{if } G_t > 0 \\ 0, & \text{otherwise} \end{cases} \qquad (4.4)$$

Then, equation (4.4) is retransformed as

$$f(G_t) = \max(0, G_t) \qquad (4.5)$$

Then, the ultimate representation of the ReLU activation function is performed using the following equation:

$$act(G_t) = \max(0, [K_t, G_t + h_t]) \qquad (4.6)$$

Translation invariance is introduced using the pooling layer to minimize layer dimensions. Varied mean and max values of feature weights are gathered from different image regions of a convoluted image. For a particular layer, the output feature map gets updated using the following equation:

$$K_{tuv} = \frac{\max}{s, r \in G_t} (E_{tsr}) \qquad (4.7)$$

where the elements of a specific image region (s, r) of an image are represented by E_{tsr} using the pooling layer, and the output feature map generated from the pooling layers is given by K_{tuv}. Then, fully linked layers are employed to extract feature vectors from the output feature map generated from the pooling layers. This layer links output of the previous layers to all the input layers. Then, the final training loss is determined by the equation:

$$F(s, r) = \frac{1}{L} \sum_{n=1}^{L} (S_n - r_n)^2 \qquad (4.8)$$

where the square variance between labels of ground truth and estimated labels is given by $F(s, r)$. Here, is the number of training images and labels of ground truth expressed by and the estimated clas

$$G_t = [G_{t1}, G_{t2}, G_{t3}, \ldots \ldots G_{tL}]^T \qquad (4.9)$$

Then, the loss function in equation (4.8) can be rewritten by substituting G_z at the place of and using equation (4.6),

$$F(S, K_t, G_t, h_t) = \frac{1}{L} \sum_{n=1}^{L} (S_n - act(G_{t_n}))^2 \qquad (4.10)$$

Then, the final loss function is given by the following equation:

$$F(S, K_t, G_t, h_t) = \frac{1}{L} \sum_{n=1}^{L} (S_n - \max(0, [K_t, G_{t_n} + h_t]))^2 \qquad (4.11)$$

4.5   PLANT DATASET

This section discusses the simulation results for plant classification and plant species identification using the proposed CNN model. The experimental results are obtained in terms of classification accuracy, precision, recall, and F1-measure. The proposed CNN model is tested on the Vietnam Plant dataset [13]. This dataset is very large and segregated into train and test images in the ratio of 60:40. The total number of training images present in the Vietnam Plant dataset is more than 12,000 and the number of testing images is more than 8000. The number of plant species present in this dataset is 200. Varied fine-grained discriminative features are observed from these training images using the proposed CNN-PC model. The size of these training images is 128 *128 pixels. The VNPlant-200 dataset images are gathered from the National Institute of Medicinal Materials in different locations, such as Ngoc Xanh Island Resort, Ho Chi Minh City, and Phu Tho City. This dataset is complex and challenging due to the presence of noise and illumination, and many leaves appear in a single image which makes their classification quite difficult. These images contain flowers, leaves, tree bark, soil, varied backgrounds, etc.

Based on these extracted dataset images, simulation results are obtained in terms of accuracy, precision, and recall using the proposed CNN-PC model. Pre-processing is performed to minimize errors and ambiguities and generate pre-trained weights from given input images from the deep analysis using varied convolutional layers, pooling layers, and dense blocks, and some essential decisions which can improve training efficiency are derived. Based on those obtained feature weights and essential parametric and functional decisions, efficient training of the Vietnam Plant dataset is performed using the proposed CNN-PC model. Finally, efficient training is performed to generate feature maps and perform classification. Classification performance is observed by comparing predicted labels using the proposed CNN model against pre-defined ground truth labels.

This section provides details about the Vietnam Plant dataset (VNPlant-200 dataset), testing performance, and comparison of obtained classification performance. The classification performance results are compared against varied plant classification models. The Vietnam plant dataset was utilized to test the performance of the proposed CNN-PC model. There are a total number of 200 classes present in this dataset which are gathered from different plant nurseries in Vietnam and the dataset contains details of varied plants like *Abelmoschus sagittifolius*, *Abrus precatorius*, *Alpinia officinarum*, *Capsicum annuum*, etc. In Figure 4.3 each row demonstrates images of some of the different classes from the Vietnam dataset.

The performance of the proposed CNN-PC model is compared against various classical classification techniques like VGG16, Inception V3, and MobileNet V2 [12] in terms of classification accuracy. All these previous models are trained on the Vietnam Plant dataset. Keras and Tensor Flow are employed as pre-trained models to generate high-quality unique features. The original ImageNet classification model is modified according to its specific goals to classify 200 plant species. Moreover, all the experiments and simulation results are performed

FIGURE 4.3 *(a) Abelmoschus sagittifolius. (b) Abrus precatorius. (c) Abutilon indicum. (d) Acanthus integrifolius.*

on an i7 processor, 16GB RAM, 2 TB SSD+HDD, and GeForce RTX NITRO5 GPU memory. Training is performed on a large image dataset and pre-defined functions of Keras are used for the enhancement of the data processing mechanism of the proposed CNN model such as image transformation, flipping, resizing, and zooming to avoid over-fitting. In this work, the generated feature maps are utilized to get testing results, and results are obtained by modifying the pre-trained ImageNet dataset.

There are a total number of 200 classes utilized. Figure 4.4 shows the suggested CNN-PC model for all 200 classes. The total number of 20,000 images is distributed into the training, validation, and testing datasets. Here, the number of training images in each class is 80, the validation dataset consists of 20 images and the testing dataset contains 40 images in each class. Thus, the training dataset consists of nearly 16000 images, testing dataset consists of almost 8000 images. Finally, the validation dataset contains a total number of 4000 images

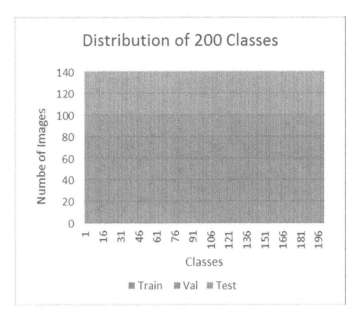

FIGURE 4.4 Precision results using the proposed CNN-PC model.

4.6 EVALUATION

Various ML techniques are evaluated, and their performance metrics are estimated to define the malicious RDP sessions.

$$Accuracy = \frac{TP + TN}{Total\ subjects} \times 100\% \tag{4.12}$$

$$Precision = \frac{TP}{TP + FP} \times 100\% \tag{4.13}$$

$$F1\ score = 2 \times \frac{TP}{TP + FN} \tag{4.14}$$

$$Sensitivity\ /\ Recall = \frac{TP}{TP + FN} \times 100\% \tag{4.15}$$

$$AP\ score = \sum_{n} \left(Recall_{n} - Recall_{n-1} \right) \times Precision \tag{4.16}$$

TP is True Positive, TN is True Negative, FP is False Positive, and FN is a False Negative value.

Classification accuracy when considering all 200 classes is 96.42%. These results are compared against the traditional classification methods like VGG16 [14], Inception

TABLE 4.1
Estimation of Performance Metrics

Classifier	Accuracy	Precision	Recall
VGC16	70%	90%	96.0%
Inspection V3	81%	92%	45%
Mobile Net V2	86%	86%	83.1%
CNN PC	96.6%	98%	99%

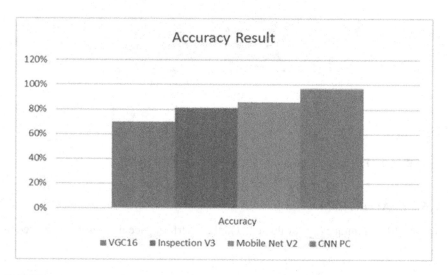

FIGURE 4.5   Classification accuracy results using the proposed CNN-PC model.

V3 [15], and MobileNet V2 [16]. The graphical representation of the classification accuracy using the proposed CNN-PC model against previous CNN classification models is presented in Table 4.1. It is evident from Figure 4.4 results that classification accuracy using the proposed CNN-PC model is comparatively higher than classical CNN models. Figure 4.5 shows the comparison of precision considering all 200 classes against various classification models, such as Resnet 50, Inception V3, and MobileNet V2. The obtained precision results using the proposed CNN-PC model is 98%. The previous best classification model in terms of precision results was Inception V3 with 92%. The proposed CNN-PC model outperforms these classical CNN classification models in terms of precision and classification accuracy.

Furthermore, Figures 4.5 and 4.6 show the graphical representation of classification performance using the proposed CNN-PC model. The mean testing accuracy, mean validation, mean precision results, and mean area under the curve (AUC) results obtained considering all 200 classes are 96.42%, 95.64%, 95.56%, and 95.47%, respectively, as shown in Figure 4.7. It is evident from classification results that the proposed CNN-PC model outperforms previous CNN classification models in terms of classification results and detects plant species accurately.

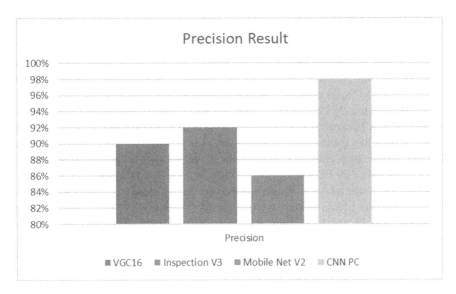

FIGURE 4.6 Classification precision results using the proposed CNN-PC model.

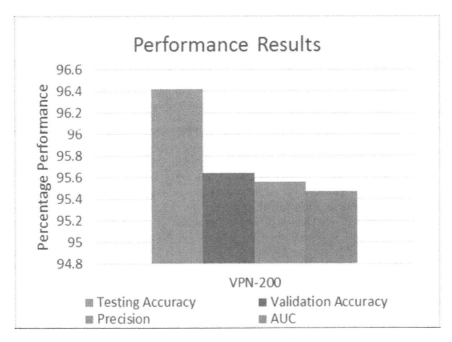

FIGURE 4.7 Performance result.

4.7 CONCLUSION

The significance of plant species detection and its classification is very essential to accurately find out whether the respective plant comes under the banner of medicinal plants or not and to define the type of plants. In this research work, a Convolution Neural Network Architecture (CNN)-based Plant Classification (CNN-PC) model is proposed to accurately identify plant type and perform classification to analyze which particular plant belongs to which species. The proposed CNN-PC model consists of several phases like the data selection and acquisition phase, the pre-processing phase, the feature generation phase, the training phase, and the testing phase. The architecture of the proposed CNN-PC model is a combination of different layers such as convolutional layers, pooling layers, soft-max layers, fully linked layers, and various blocks like convolutional blocks and dense blocks. A deep mathematical representation for the proposed CNN-PC model is discussed. Test performance results using the proposed model are determined based on the Vietnam Plant dataset (VNP-200). Here, the number of plant species within the VNP-200 dataset used to test the performance is 200. The obtained mean testing accuracy considering all the 200 classes is 96.42%. Furthermore, mean validation accuracy and mean precision considering all the classes are 95.64% and 95.56%, respectively. The classification results confirm that the proposed CNN-PC model outperforms the previous CNN classification models significantly.

REFERENCES

[1]   Fabricant, D. S., and Farnsworth, N. R. "The value of plants used in traditional medicine for drug discovery," *Environ. Health Perspectives*, 109, no. 1, pp. 69–75, 2001.

[2]   Bennett, B. C., and Balick, M. J. "Does the name really matter? The importance of botanical nomenclature and plant taxonomy in biomedical research," *J. Ethnopharmacol.*, 152, no. 3, pp. 387–392, Mar. 2014.

[3]   Rivera, D., Allkin, R., Obón, C., Alcaraz, F., Verpoorte, R., and Heinrich, M. "What is in a name? The need for accurate scientific nomenclature for plants," *J. Ethnopharmacol.*, 152, no. 3, pp. 393–402, Mar. 2014.

[4]   Bussmann, R. W. "Taxonomy_an Irreplaceable Tool for Validation of Herbal Medicine," in *Evidence-Based Validation of Herbal Medicine*, pp. 87–118. Amsterdam, The Netherlands: Elsevier, 2015.

[5]   Bennett, B. C., and Balick, M. J. "Phytomedicine 101: Plant taxonomy forpreclinical and clinical medicinal plant researchers," *J. Soc. IntegrativeOncol.*, 6, no. 4, pp. 150, 2008.

[6]   Wu, Z., Chen, Y., Zhao, B., Kang, X., and Ding, Y. "Review of weed detection methods based on computer vision," *Sensors*, 21, no. 11, pp. 3647, May 2021.

[7]   Li, Z., Guo, R., Li, M., Chen, Y., and Li, G. "A review of computer vision technologies for plant phenotyping," *Comput. Electron. Agricult.*, 176, Sep. 2020, Art. no. 105672.

[8]   Haque, Md., and Adolphs, Julian. (2021). Corn leaf disease classification and detection using deep convolutional neural network. https://doi.org/10.13140/RG.2.2.20819.50722

[9]   Gajjar, V. K., Nambisan, A. K., and Kosbar, K. L. "Plant identification in a combined-imbalanced leaf dataset," in *IEEE Access*, 10, pp. 37882–37891, 2022. doi: 10.1109/ACCESS.2022.3165583

[10] Zhang, Y., Wa, S., Liu, Y., Zhou, X., Sun, P., and Ma, Q. High-accuracy detection of maize leaf diseases CNN based on multi-pathway activation function module. *Remote Sens.*, 13, pp. 4218, 2021. https://doi.org/10.3390/rs13214218

[11] Srilatha, Doddi, and Thillaiarasu, N. "Implementation of intrusion detection and prevention with deep learning in cloud computing." *J. Inf. Technol. Manag.*, 15, Special Issue, pp. 1–18, 2023.

[12] Shyamambika, N., and Thillaiarasu, N. "A Survey on Acquiring Integrity of Shared Data with Effective User Termination in the Cloud." *2016 10th International Conference on Intelligent Systems and Control (ISCO)*. IEEE, 2016.

[13] Thillaiarasu, N., and ChenthurPandian, S. "Enforcing Security and Privacy over Multi-Cloud Framework Using Assessment Techniques." *2016 10th International Conference on Intelligent Systems and Control (ISCO)*. IEEE, 2016.

[14] Preethi, P., Asokan, R., Thillaiarasu, N., and Saravanan, T. "An effective digit recognition model using enhanced convolutional neural network based chaotic grey wolf optimization." *J. Intell. Fuzzy Syst.*, 41(2), 3727–3737, 2021.

[15] Thillaiarasu, N., Pandian, S. C., Vijayakumar, V., Prabaharan, S., Ravi, L., and Subramaniyaswamy, V. "Designing a trivial information relaying scheme for assuring safety in mobile cloud computing environment." *Wireless Networks*, 27, 5477–5490, 2021.

[16] Shenoy, Ashwin, and Thillaiarasu, N. "A Survey on Different Computer Vision Based Human Activity Recognition for Surveillance Applications." *2022 6th International Conference on Computing Methodologies and Communication (ICCMC)*. IEEE, 2022.

Big Data Analytics

5 Big Data Visualizing with Augmented and Virtual Reality

Challenges and Research Agenda

*G. Bindu Madhavi, Y. Sowmya Reddy,
G. Ravi Kumar, and Ajmeera Kiran*

5.1 INTRODUCTION

All of humankind's recorded past constitutes a massive database. Data storage has been around for millennia. History, politics, science, economics, business, and even social life have all grown to rely heavily on data. This pattern is most evident in social media platforms like Facebook, Twitter, and Instagram, where users produce a deluge of content every day [1]. The public can now access and utilize data from government agencies, scientific and technical labs, and space research projects. For instance, the 260 gigabytes of human genome data made available by the 1000 Genomes Project [2] is just one example. The Internet Archive and ClueWeb09 are just two of many places where you may access more than 20 terabytes of data for free. In recent years, technological advancements have facilitated greater awareness and responsiveness to our immediate environments. Decision support systems (DSS) are one such tool, since they are used to facilitate the DSS in a wide range of contexts. Information science, cognitive psychology, and Artificial Intelligence (AI) are just a few of the fields that have actively pursued study into these systems. Several scientific disciplines have dedicated extensive research and development resources to improving the decision-making process. When faced with a variety of options, "good decision-making" occurs when people are well-informed and armed with all the data they need to make an educated call [3]. Companies may now process Big Data with fewer resources and at a lower cost. Big data processing is rapidly gaining traction in the commercial world because the benefits outweigh the investment required [4]. International Data Corporation (IDC) claims that data trading is developing into its own industry. Seventy percent of large companies already buy external data, and that number is predicted to rise to 100% by the beginning of 2025.

DOI: 10.1201/9781003388241-7

5.1.1 BIG DATA

Every day, businesses and social media platforms produce reams of data, which are typically represented in web blogs, text documents, or machine code, such as geo-spatial data that may be gathered in diverse stores both inside and outside of an institute. The usage of data centers, cloud storage, and distributed data repositories is also extremely widespread. Additionally to the practice of creating the foundation for momentous conclusions, businesses now have the tools they want to establish the association amid data parts. With the ever-increasing speed at which data can be processed, the possibility of a situation arising in which conventional analytical methods would be unable to keep up arises, paving the way for Big Data technology [5]. In this chapter, you'll learn about the many forms of existing data that may be analyzed with the use of specific methods. Many new visualization techniques have emerged recently for the efficient depiction of preprocessed data. Multi-dimensional volumetric visualizations have replaced two-dimensional flat graphics. New methodologies bring forth new research issues and solutions, both of which will be covered in the paper that follows. Nonetheless, the evolution of Big Data visualization cannot be regarded as complete [6].

5.1.2 BIG DATA PROCESSING METHODS

Numerous methods exist today for analyzing data, most of which are founded on statistical and computational methods. ANN, models dependents on the theory of the design and operation of biological NNs, methods of predictive-analysis, statistics, Natural Language Processing, etc., are some of the most cutting-edge approaches to analyzing massive datasets. Methods for dealing out Big Data draw from a wide range of academic fields, such as economics, computer science, statistics, and applied mathematics. Data Mining, NNs, ML, signal-processing, and visualization methods all have them as their foundations [7]. System utilization is dramatically improved as a result of the interconnectedness and simultaneous use of most of these approaches during data processing, as shown in Figure 5.1.

Designers, programmers, and scientists are all on the lookout for innovative visualization methods and prospects right now. Business decisions at Amazon, Twitter, Apple, Facebook, and Google, among others, are aided by data visualization [8]. Insights from multiple points of view can be gained via visualization tools for business. As a first step toward bettering the company's ties with its customers, the deployment of cutting-edge visualization tools enables the quick study of all customers'/users' data. Marketers can use this information to divide their target audience into more targeted groups based on things like their interests, demographics, and past purchases. The most lucrative customers or users can be identified and studied with the use of correlation mapping. Second, businesses can take advantage of opportunities to discover connections between product, sales, and customer profiles thanks to visualization tools. Companies might use the data they collect to provide interesting new offers to their clientele. In addition to helping with risk analysis, visualization allows for the monitoring of revenue changes. As a third benefit, data visualization facilitates deeper comprehension [9]. Obtaining information that is relevant, consistent, and accurate

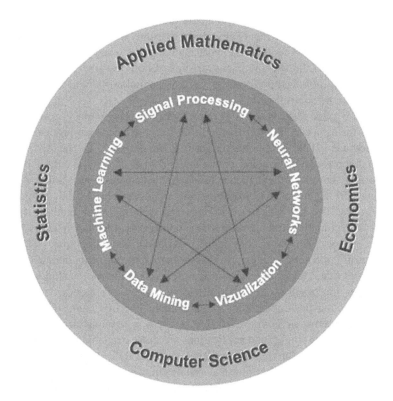

FIGURE 5.1 Big Data processing methods interconnection.

leads to greater efficiency. Thus, data visualization may help businesses explore alternative profitable advertising strategies. Here, we introduced the reader to the most common data analysis methods and elaborated on the close relationship between them. However, the Big Data era has only just begun to develop. As a result, new Big Data processing techniques are constantly being created to address the challenges of Big Data. By this, we mean that it is necessary to employ a wide variety of multidisciplinary approaches to Big Data in order to fully comprehend its intricate structures and relationships [10].

5.1.3 AUGMENTED REALITY

Learning performance, motivating factors, and attitudes are just a few of the areas where studies have demonstrated that augmented reality (AR), the visualization of virtual in combination with physical, real-world information, can improve outcomes. Different AR applications were utilized to support teaching of various topics and achieve varying learning goals in the studies reviewed here [11]. Despite this, the research often centered on field experiments where an AR application was compared to a traditional learning setting without noting the exact characteristics that may lead to its superiority, suggesting that AR may be effective in a variety of learning contexts.

Studies comparing different types of media are often criticized for a number of reasons, including the following: (1) the medium is only a means to an end; the information and instructional method are what really matter when it comes to students' success in class; (2) the specific characteristics of both the students and the media that contribute to improved performance in class are unknown; and (3) confounding variables cannot be controlled for [12].

Industry 4.0, often identified as the Fourth Industrial Revolution, is now converting the manufacturing sector by prioritizing the development of smart products and production methods. There is a contemporary driving force at the heart of industry expansion that represents the achievement of large-scale changes in current industrial manufacturing: the promise of higher productivity and flexibility. Smart manufacturing and smart maintenance technologies and processes are essential to overcoming these obstacles. The IoT and other systems and techniques for Industry 4.0 is an approach to meeting the growing need for versatility and efficiency in manufacturing that makes use of cyber-physical systems, Big Data, cloud-computing, AR, and VR [13]. The term "internet of things" (IoT) refers to a network of interconnected computing devices that enables diverse physical items like sensors, machines, vehicles, and structures to communicate with one another and work together to achieve a shared purpose. IoT has implications in many areas. The IIoT (Industrial Internet of Things) is the Internet of Things used in a business or manufacturing setting. New interconnection technologies and applications are made available to internet-enabled cyber-physical systems through the implementation of IIoT. With this point of view, process data on industrial cyber-physical systems may be transformed, analyzed, and controlled in real time across several locations. Interoperability, virtualization, decentralization, modularity, and real-time data collection and analysis are just some of the smart manufacturing and maintenance methods and systems that have seen a surge in popularity over the past few years [14].

Surry and Ensminger [15] suggest conducting research that focuses on the medium itself (intra-medium studies) and the learner (aptitude-treatment-interaction studies) to better understand how the medium affects each. These methodological approaches to study can also shed light on the complex interplay between the features of the learner and the characteristics and variables of AR presentations and experiences and the consequences of the learning process. The knowledge obtained can aid educators and developers in determining when and how AR can best serve their students' educational needs. In this paper, we implement an intra-medium comparative study with an aptitude-treatment-interaction in response to Surry's and Ensminger's recommendations for alternative research [16].

Scalability of unstructured data, availability, real-time analysis, and fault tolerance are only some of the difficulties associated with big data. Big data analytics benefits from a number of cutting-edge technologies, including collaboration on the cloud, machine learning, and immersive realities like augmented and virtual reality (AR/VR). One of the best ways to comprehend intricate elements of big data is through the use of information in the form of virtual graphics. The human mind has been conditioned to retain more information if a relevant image is associated with it. Industry 4.0, which relies heavily on analytics, benefits from the development of big data technologies. However, a company's success may hinge on its ability to store massive amounts of data [17]. AR/VR offer consumers a synthetic representation of their surroundings. Immersing the user in a digital environment, this technology produces and expands reality with potential applications in fields as varied as transportation, healthcare, and industry. Virtual reality, in which a computer alters a user's perception of the real world, serves as the theoretical foundation for augmented reality. AR is a technique that superimposes computer-generated augmentation on top of real-world environments, while VR necessitates a totally virtual environment [18]. The duties are completed more effectively and efficiently through AR. It can be thought of as a hybrid of the actual world and virtual reality. Using AR, we can render data from the actual world and present it in a way that makes a virtual element look just like a real one.

In order to show information virtually in a real-time setting, augmented reality (AR) relies on big data as its foundation. Various ways exist for visualizing data and its attributes, including 2D and 3D graphics, geometric transformation, display icons, pixel-based and hierarchical-based approaches, tag clouds, clustergrams, motion charts, and dashboards. The above techniques can be used to visualize large datasets [19]. Visual data is being used by many businesses as a marketing tool as the globe braces itself for the onslaught of COVID-19. Push and fetch operations from cloud storage have become feasible thanks to lightning-fast Internet connections. Remote services are useful in many settings, including businesses, hospitals, and educational institutions. With augmented reality, you can virtually experience anything, even if it is not physically present. In order to better explain their product to customers, several businesses are turning to virtual and augmented reality technologies. Consumers won't need to be physically present to benefit from its services because they can experience them virtually instead. QoS can also be raised by enabling the display of huge data [20]. Visualization may help project numerous practical aspects and provide examples with more effect in the field of teaching. The use of augmented reality techniques like these helps students better understand complex ideas. In the COVID-19 scenario, college students learn independently. Therefore, they can use visualization to simulate being in a lecture hall or laboratory. During the height of the COVID-19 outbreak, when over a third of the planet was on lockdown, corporations and offices had only a limited number of options for carrying out their daily operations. Visualization was one approach that was very useful to companies. Since most individuals nowadays live in relative isolation, with all services being wholly reliant on data generation and processing, the concept of visualizing big data has proven useful in the medical and banking sectors as well [21].

Tools that facilitate the efficient and rapid processing of huge datasets are the focus of current research in the subject of Big Data visualization. Additionally, the analysis of the presented data might be evaluated from every conceivable perspective in unique, scalable methods [22]. We summarize the main difficulties in visualizing Big Data from the existing literature on the topic and suggest a new technical method to visualizing Big Data by combining insights into human perception with cutting-edge Mixed Reality (MR) technologies. Our research suggests that combining existing Big Data visualization methods with AR and VR, which are optimized for human perception limitations, is a potential approach to improving upon them. We outline the critical next steps for the research outline to put this strategy into action.

This chapter touches on a wide range of topics, although the study is headed in one of three broad trajectories.

- The limitations of the human brain in Big Data visualization.
- Opportunities in AR/VR for Big Data visualization.
- What are the advantages and disadvantages of the proposed visual method?

This chapter's structure continues as follows. In section 5.2, we will discuss what AR and situated visualization (SV) are. Section 5.3 then offers augmented and virtual reality integration, section 5.4 discusses benefits, obstacles, and research prospects, and section 5.5 provides a discussion on the future research agenda and the major problems of data visualization. The chapter is concluded with a summary in section 5.6.

5.2 AUGMENTED REALITY (AR)-BASED VISUALIZATION AND SITUATED VISUALIZATION

5.2.1 AUGMENTED REALITY VISUALIZATION

The term "human-machine interaction tools that overlays computer-generated information in the real-world environment" is a good summary of the basic idea behind augmented reality. It is the seen things themselves that set the stage for the exposition and overlay of information. This definition is intended to be interpreted broadly. The possibility exists for AR to be used in conjunction with other senses, displaying data that is not normally accessible to or observable by the human senses. Figure 5.2 depicts the three essential features necessary for an AR experience, as stated by Azuma [23]. These characteristics are the integration of actual and virtual content, real-time interaction, and 3D registration.

FIGURE 5.2   Sample of visualization in context utilizing AR.

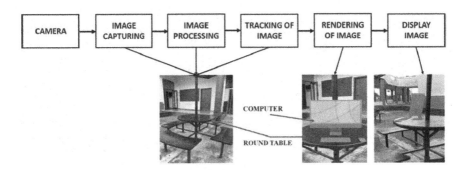

FIGURE 5.3 Concept diagram for AR visualization.

The term "augmented reality" can be defined in two ways. Augmented reality and augmented virtuality occupy a place on the "virtuality-reality" continuum first proposed by Paul Milgram in 1994. In a mixed reality, both the actual world and the virtual world are fused, with AR based more on the former and VR on the latter. Figure 5.3 depicts this concept of augmented reality, which occurs when computer-generated imagery is superimposed on a user's view of the real world [24].

In the reality-virtuality continuum postulated by Reddy et al. [25], augmented reality could be seen as a transitional stage between VR (showing a virtual world) and the unaltered actual world. The purpose of both VR and AR is to immerse the user, yet these two technologies take quite different approaches. AR employs computer-generated equipment to merge virtual reality with real life, displaying virtual elements as an overlay to the actual world, making it more relevant through the ability to interact with existing virtual aspects, while VR provides a digital replica of a real-world setting. By engaging with them, you may gain a more nuanced understanding of the world around you. Unlike virtual reality, augmented reality (AR) does not seek to completely replace the physical environment but rather to deliver virtual stimuli while maintaining the sense of presence from the individual experiencing it. Two substitute references may be utilized by AR, depending on the circumstances. The most common technique for determining when and where to display virtual content is to use visual clues/labels to supplement existing knowledge about physical parts. The second method, known as location-based, utilizes technologies like global positioning systems (GPS) and other sensors to estimate the user's position and then displays content accordingly. Although not AR under Azuma's definition, the term is commonly used to describe a technique that displays information that is not spatially aligned with physical objects [26].

Data visualization is a means of information dissemination, defined as the transformation of obscure information into readily digestible visuals. According to Munzner [27], "computer-based visualization systems provide visual representations of data sets designed to help people carry out tasks more effectively," and "visualization is suitable when there is a need to augment human capabilities rather than replace people with computational decision-making methods." There are two main categories of AR visualizations: visual AR (VAR) and spatial AR (SAR). As may be seen in

FIGURE 5.4   Instances of AR visualizations: (a) Visual AR. (b) Spatial AR.

Figure 5.4a, VAR involves the superimposition of computer-generated content onto the user's field of view. As can be seen in Figure 5.4b, in SAR the digital information is superimposed on the real world.

5.2.2 SITUATED VISUALIZATION

Because augmented reality is not tied to a single location, mobile and outdoor solutions are a natural fit, enabling engagement in the environment through a variety of tracking methods (including markers, sensors, GPS, and user interfaces). Situated visualization, which refers to visualization in relation to its situation, builds on this idea. One of the key benefits of AR systems is that extra digital data of the route may be displayed and studied directly overlaid on photos of that environment, which is impossible without the use of AR [27]. Situated visualization (SV) is a tool introduced to provide a more detailed understanding of a situation. It takes into account visualizations that are pertinent to the actual environment in which they are exhibited, and so includes all visualizations that alter their appearance depending on that environment. What White meant by "through the combination of the visualization and the relationship between the visualization and the environment" can be summed up as "when the visualization of the virtual information is intrinsically related to its environment." Figure 5.5a shows the amount of concentration and location of carbon dioxide on a roadway, received from an environmental sensor, which are examples of SV based on AR applications that depict the subterranean infrastructure of the place where the user is. Figure 5.5b shows how to use a personal interaction panel to identify locations of interest in city settings, and Figure 5.5c shows how to use a panel to examine common aspects of dynamical systems [28].

5.2.3 CHALLENGES WITH AUGMENTED REALITY IMPLEMENTATION

While several sectors are quickly adopting augmented reality, users still face certain fundamental obstacles. The following are just a few examples of these difficulties:

- Users were hesitant to adopt AR-ready systems due to their bulkiness. Because of their limited computing power, HMDs are often compared to laptop computers.

FIGURE 5.5 Instances of: (a) situated visualization (SV) of physically-based data type; (b) SV of abstract data type; (c) non-SV.

- Heavy software and hardware are needed; therefore it is no surprise that data processing is often interrupted.
- Improvements are needed in the user interface and tracking mechanism of augmented reality components in order to enable accurate and rapid tracking.
- Why most people aren't familiar with AR and VR devices, so teaching them how to use them takes a long time and requires extensive training.
- Privacy and data security policies must be strengthened.
- The augmented reality system must be able to communicate effectively with other systems.
- The AR systems are extremely perceptive; this sensitivity results in a decrease in processing speed when interruptions are present.

5.3 DATA VISUALIZATION METHODS

5.3.1 CLASSIFICATION

The vast volumes of abstract data can be transformed into representations that are useful for perception, cognition, and communication using one of the many computer visualization tools available today. This categorization takes into account the tool's features from the user's point of view, including their level of expertise with the tool, its capacity to facilitate teamwork, its adaptability, and its scalability. This study does not account for the presence of premade templates, interactive visualization, or other crucial features. The proposed interfaces' ability to support several languages eliminates the need to categorize instrument makers as either domestic or international [29].

5.3.2 ANALYSIS OF BIG DATA VISUALIZATION APPROACHES

The decision-maker needs an accurate representation of the data collected from the IoT network as a whole. Visualization is an effective method for these uses. Attributes and variables for a data point are shown systematically in data visualization. Tables, charts, pictures, and other user-friendly representations of data are all products of data visualization techniques. Data visualization for large datasets is more complex than for smaller datasets [30]. Traditional methods of visualizing data have begun to grow, but this is insufficient. Before actually visualizing massive amounts of data, several researchers perform feature extraction and geometric modeling to drastically minimize the data size. When viewing large amounts of data, it is crucial to use the appropriate data representation [31]. Common data visualization techniques include the use of tables, histograms, scatter plots, line diagrams, pie charts, block diagrams, bubble charts, hybrid charts, time lines, entity connection diagrams, etc. In addition to the aforementioned techniques, treemaps, parallel coordinates, cone-shaped trees, semantic networks, etc. also fall under the umbrella of "standard methods of data visualization." Data items in several dimensions are built using parallel coordinates (Figure 5.6). When dealing with multidimensional data, this form of data visualization comes in quite handy [32].

The main challenges in visual analytics are related to the scalability and dynamism of large data. The complexity of large data visualization stems from the wide variety of data types involved (structured, semi-structured, and unstructured). As was mentioned before, velocity (speed) is crucial for studying Big Data. Combining cloud-computing and a high-end GUI with large data can improve scalability management [33]. Due to bandwidth constraints and power needs, the visualization must conform to the data in order to efficiently obtain current information from big data, which is often in an unstructured data format. The sheer volume of information necessitates the employment of parallel calculations: the process of allocating work among multiple processors so that multiple tasks can be worked on in parallel [34]. Big data visualization has the following issues: there is a lot of visual clutter; the majority of the dataset's objects are too closely related to one another; users are unable to visually distinguish them as distinct elements on the screen, leading to data loss and complicated picture perception; simplified calculations through data reduction, which can result in lost information. Data visualization techniques are constrained not just by the viewing device's aspect ratio and resolution, but also by the viewer's own sensory capabilities; users watch the data but have no control over the volume or brightness of the changes displayed; and stringent performance standards [35].

In contrast to the benefits of a holistic view of the data, a reduction in data by sampling or filtering might result in the loss of interesting emissions and deviations without sacrificing the schedule's readability or the users' perceptions and cognitive abilities. High latency, which hinders real-time engagement, might result from requesting massive data warehouses [36].

Solutions for Big Data visualization issues:

- Having one's demand for velocity met. This necessitates more memory and robust parallel processing from the hardware side.
- Realizing what the data means. One answer is for those who receive the data to have sufficient prior knowledge and experience to quickly grasp its meaning.
- Resolving issues with the reliability of data. It is important to make sure that any extraneous information is removed from the data before it is sent to the visualization.
- One approach is to pool the information into a bird's-eye view, from which subsets of the whole can be seen and the data can be displayed more clearly.

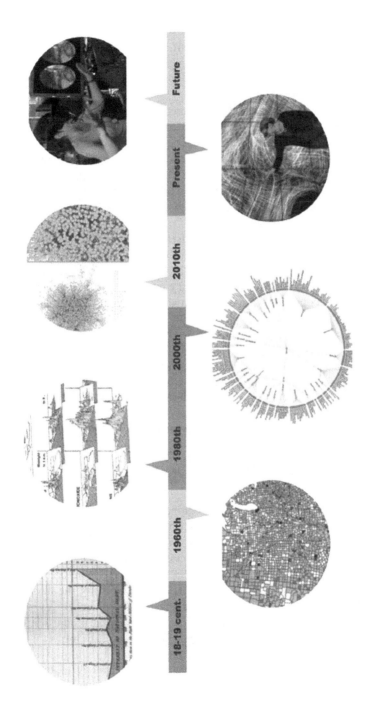

FIGURE 5.6   The evolution of visualization methodology.

- Make some allowances for emission cases. One approach could be to filter out unusual data points or to plot them on a separate chart.

Data criteria such as massive data support, support for multiple types of data, and the capability to vigorously display data changes were used to categorize and examine the various approaches and methods for visualizing big data.

- Both Treemap and its direct competitor, Circle Packing, are methods for representing hierarchical data visually. However, it employs circles as its base shape, and higher-level circles can be incorporated into lower-level ones.
- Sunburst is a technique that transforms treemap viewing into a polar coordinate system. The most notable distinction is that the arc's radius and length are the varying parameters, rather than the arc's breadth and height.
- Parallel coordinates, which broadens visual analysis by incorporating many data components per object type.
- Streamgraphs are a special kind of offset graph in which the stack region is rotated around a central axis, producing a curvy, organic shape.
- Data objects are arranged in a circular pattern, with arcs connecting them according to the strength of their ties in a circular network diagram. The consistency of an object is typically evaluated by comparing its line width or color saturation to others.

Hadoop is compatible with a wide variety of data visualization tools. Hadoop MapReduce, Hadoop HDFS, Hadoop YARN, and Hadoop Common are all examples of Hadoop components. Separate software exist, but they all display data in a two-dimensional manner, making it hard for the user to comprehend. Using augmented reality as a data visualization tool is the answer, as it will make the user's life easier and present the information in a more manageable format.

5.3.3 BIG DATA, IoT, AND AR: TECHNOLOGY CONVERGENCE IN VISUALIZATION ISSUES

In 1979, the United States experienced a significant nuclear power plant catastrophe due to human error brought on by workers who misinterpreted incoming data and failed to spot the looming crisis in time [37]. Work in many fields will be made safer and more efficient with the aid of AR in the visualization of data obtained through Internet of Things (IoT) sensors. The sensors' readings can be monitored in real time, the data may be produced graphically, and the user can control the device from a distance, all thanks to this technology. With the help of the smart sensors installed in every area, the manager will be able to extra exactly control the production situation, and the workers who are closest to the equipment will be able to inspect it more frequently. If there is an issue, employees can either correct it right away or send the data to engineers outside the facility who can help. Technology like this also has applications in education and teaching. Interactive 3D representations of equipment and visual instructions can be incorporated into augmented reality manuals that businesses can create. Caterpillar Inc., a multinational company, uses AR effectively to educate and empower its engineers. The fields of medicine, teaching, aerospace, tourism, and property all offer numerous instances [38].

The volume of augmented reality processing has a direct effect on the device's power consumption and lifespan, and the connection between AR and IoT necessitates a significant number of calculations. Therefore, an additional cloud database is required to work with augmented reality on smartphones in order to lighten the load, conserve energy, and circumvent the phone's memory constraints [39]. Since moving all computing activities to the cloud platform might cause significant delays in the operation of the program, it is important to note that the usage of cloud technologies should be governed by adaptive management. Keeping an eye on cloud server load and network congestion is essential for optimizing cloud usage. The health of the network, the availability of the cloud server, the battery life of the device, and the user's impression of the experience (QoE) must all be constantly tracked. Workers can keep an eye on connected equipment, diagnose issues, and perform maintenance using augmented reality apps on smartphones, tablets, or enterprise AR headsets [40]. Enterprises can benefit from augmented reality (AR) headsets like Google Glass Enterprise and Microsoft HoloLens by providing step-by-step online training and tutorials for building and operating equipment. Existing augmented reality software development kits (SDKs) like ARToolkit, EasyAR, OpenCV, Maxst, Kudan, Catchoom On-Device Image Recognition SDK, and Vuforia SDK can be used to construct an AR app that can display data from IoT sensors. Such SDKs allow for the creation of an AR component, but additional applications are required to link IoT devices and collect data [41].

There are pre-built systems that incorporate data visualization and sensor connectivity, two essential components of every successful program's deployment. PTC, Augmenta, Bosh, and Reflect are the most popular platforms for making augmented reality applications for the Internet of Things. The real-time viewing of massive data from IoT devices on mobile devices with augmented reality functionality is now the key area of convergence of the listed technologies. Commonly, we can categorize convergent phenomena as follows:

1. Sensor data visualization and object administration.
2. The use of decision support systems and equipment performance monitoring. Timely troubleshooting of equipment issues and enhanced product quality are made possible by the data's visual representation. In an ideal world, this technology would be used throughout the entire manufacturing process.
3. Since mobile devices (smartphones, tablets, smartwatches, and augmented reality helmets) are screens for data visualization, the subject of how much power they consume has recently come into the spotlight.

5.4 AUGMENTED REALITY FOR BIG DATA

In order to maximize the value, volume, velocity, and diversity of big data, it is necessary to use 5G wireless networks. Increasing network capacity necessitates implementing measures like increasing network density and optimizing spectrum use. When the need for gathering, processing, and delivering data enters the scene, time is of the essence. Efficient network functioning is required to support various types of data. One approach that can accommodate pace and variety is end-to-end networking. Sensors, cameras, mobile phones, drones, etc., will produce a deluge of information [42]. It is anticipated that the data rate will increase to greater than 10 Gbps with the advent of 5G. The interconnectivity afforded by the IoT is remarkable. Big Data makes it possible to perform the arduous work of integrating data from multiple sources. With 5G, gadgets will be able to talk to one another, resolving the IoT's connectivity issue [43].

Algorithms like priority scheduling and others aid in network architecture. Mission-critical services likewise rely heavily on Big data to be a success. With the help of software-defined networking, the entire network may be managed from a single location. It will reduce operating expenses for the system. The increased connectivity and convenience provided by 5G's use cases are two of the benefits it offers to users. Ultra-reliable low-latency communication (URLLC), enhanced broadband communication (EMBB), and massive machine-type communication (mMTC) all provide users with additional benefits. All of these applications rely on Big Data's enhanced connection for optimal utilization. The retail, healthcare, transportation, robotics, information technology, educational, and other industries [44] that use augmented reality are diverse.

5.4.1 RETAIL AND BANKING

The use of augmented reality is opening up new doors for the retail industry. It is useful for showcasing the product's potential. Today, augmented reality banking and retail services are being used by both major and small banks. As the number of digital clients continues to rise, augmented reality will become increasingly important. Mobile devices and social media platforms are used by these digital consumers for everything from billing and payments to product recommendations. Therefore, a massive amount of information will be utilized. Both the IoT and massive amounts of data play significant roles. The user experience in blockchain, VR, and AR is also excellent. Here, security is an issue, and striking a balance is of the utmost importance. A hundred servers are needed for the number of outlets and thousands of AR/VR machines are needed for an industrial Big Data with Augmented Reality unit. Big Data will aid the Internet of Things in handling massive amounts of data. The primary objective is to guarantee the safety of all financial dealings. 5G's fast data transfer rates make it a good fit for this type of application [45].

5.4.2 HEALTHCARE

One of the prerequisites for healthcare is instant access to the pertinent information and the ability to act on it. With augmented reality, pertinent data can be displayed instantaneously, as needed. Its use dates back at least a decade. Accessible and massive amounts of data are essential to the success of any augmented reality infrastructure. Self-tracking health data is made possible with AR as shown in Figure 5.7, which is useful for keeping tabs on health statistics. It monitors your health and shows you real-time notifications. Big data services can be utilized to keep tabs on a person's health no matter where they are, but a local storage device will also be necessary [46].

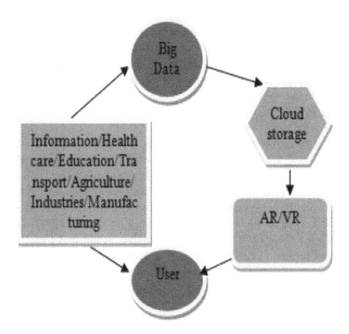

FIGURE 5.7   Big data association with AR.

5.4.3 Industry

In the current situation, where COVID-19 has had a devastating effect on everything, augmented reality and big data have played a significant role. It offers a number of benefits. From a technology standpoint, augmented reality has the potential to enhance several aspects of industry 4.0 in a variety of sectors. The idea of working remotely has been made possible by advances in technology. The augmented reality app can help maximize productivity. This pandemic has had a devastating impact on numerous industries, including those related to information, healthcare, agriculture, transportation, and education. The ability to show and visualize the items has been made possible through augmented reality. Product, building, and material visualizations can be shown at any time and in any place, whether it is an office or a storefront. The availability of "big data" allows for massive data manipulation.

5.5   ISSUES AND CHALLENGES

It is extremely unlikely that we'll ever be able to achieve better quality by completely eliminating all obstacles. Both big data and augmented reality have a number of obstacles that prevent them from reaching their full potential and satisfying their customers. When dealing with big data, the sheer volume of data is the first major hurdle. Massive amounts of data connected with either humans or machines call for a rise in the amount of storage space available, which in turn calls for ongoing analysis and improvements. It takes effort to arrange data in a sequential fashion that is fast enough. When new data isn't actively processed, it accumulates because of inadequate storage space [47]. Existing storage systems need to be modernized so that high-velocity data can be stored and accessed with ease. For maximum accuracy, it is best to go with a modern, high-performance database. The inability to systematically accommodate and file data due to a lack of awareness of data allotropes is another problem. Due to the system's lack of domain-specific knowledge, variety is a key challenge. This is because the system has not been updated to accommodate rapidly developing technologies [48].

Parameter analysis with AR is hindered by the lack of dedicated application-embedded hardware. Until now, smartphones have been the exclusive delivery mechanism for most AR apps and services. Measurement of critical object properties using application-embedded hardware other than smartphones remains the primary challenge. The door to understanding the items more thoroughly is closed as a result. Another problem appears when the circumstances are reversed [49]. The necessary components are already in place, but software that would make use of them is in short supply. There is a growing call for international standardization of holo-lens and other headsets that provide a three-dimensional image. The worldwide standardization process cannot proceed until effective apps supporting the hardware are centralized. Since AR is a relatively new concept, it is important that its consumers understand how it operates. In addition, it is necessary to create a machine learning-based system that can combine the two processes [50].

Because of the users' lack of familiarity with the technology's utility and routine use, it is crucial that it be made more accessible to the general population. Both AR and large amounts of data have similar concerns, such as privacy and security. These problems are widespread and affect all types of communication networks. Secure logins are essential for privileged access. In the age of the hack, private information, also known as personalized access, is never safe [51]. Although many different encryption methods have been used, none of them can guarantee complete anonymity around the globe. Big data and augmented reality can only coexist if both parties are able to understand the variety of data formats used in conversations between them [52]. For efficient data type conversion, having access to a high-speed converter is recommended. Since each system prefers to exchange information in its own unique format, this is strongly suggested.

5.5.1 FUTURE RESEARCH AGENDA AND DATA VISUALIZATION CHALLENGES

The average user's comprehension of the given data is greatly enhanced when visualized. People use their eyes to learn about their surroundings from the moment they are born. Images are often simpler to understand than words alone. The shift in emphasis from text-based to image-based data display and user experience is a hallmark of the modern era. In addition, visualization tools are now widely available to the general public. Now more than ever, the function of visualization when dealing with vast amounts of data needs to be reevaluated, as visual items proliferate everywhere from social media to scientific articles.

The implementation of an MR technique has many advantages, the greatest of which is the enhancement of the human experience. Simultaneously, this type of representation makes it simple to gain insight from vast amounts of data from multiple perspectives. Using both physical and verbal cues, navigation is intuitive and easy. It makes visualization effective in conveying knowledge to the end user and reduces the potential for erroneous conclusions to be drawn from data analysis. Additionally, it ensures insights that can be put into action, which boosts the quality of decisions. In conclusion, the difficulties of data visualization in augmented and virtual reality are tied to both technological progress and concerns for users. Surprisingly, several academics are already attempting to combine such diverse but interconnected areas as big data analysis, its presentation, and the complicated control of the visualized environment. It is important to remember that all of these considerations need to be made at once if the developed industrial field is to have the finest possible outcome.

5.6 CONCLUSION

In actuality, Big Data processing and analysis face many obstacles. The current state of data visualization presents challenges for data extraction, perception, and understanding. These activities are labor-intensive and not certain to yield desirable outcomes. We have produced a classification of useful Big Data visualization approaches and made some suggestions about the current trend towards visualization-based solutions for supporting business and other important disciplines. Data visualization's history and current condition were outlined, and their merits and flaws were broken down. We offer a method for utilizing VR, AR, and MR for Big Data visualization and analyze its benefits, drawbacks, and prospective optimization options. Understanding the challenges associated with human perception and restricted cognition is crucial for solving the visualization problems mentioned in this study. The design industry won't be able to help make Big Data more efficient and helpful until that time has passed. The most natural interaction with visible virtual objects and consideration of key cognitive psychology concepts may lead to improved data visualization methods. Adding features to filter out areas of reduced or no vision would also greatly enhance recognition times for those with such conditions. A move toward wireless solutions would also increase processing and quality while extending battery life.

REFERENCES

[1] Satriadi, K.A., Smiley, J., Ens, B., Cordeil, M., Czauderna, T., Lee, B., Yang, Y., Dwyer, T. and Jenny, B., 2022, April. Tangible Globes for Data Visualisation in Augmented Reality. In *Proceedings of the 2022 CHI Conference on Human Factors in Computing Systems* (pp. 1–16). CHI.

[2] Popescu, G.H., Valaskova, K. and Horak, J., 2022. Augmented reality shopping experiences, retail business analytics, and machine vision algorithms in the virtual economy of the meta verse. *Journal of Self-Governance and Management Economics*, 10(2), pp. 67–81.

[3] Adams, D., 2022. Virtual retail in the meta verse: Customer behavior analytics, extended reality technologies, and immersive visualization systems. *Linguistic and Philosophical Investigations*, 21, pp. 73–88.

[4] Slocum, T.A., McMaster, R.B., Kessler, F.C. and Howard, H.H., 2022. *Thematic Cartography and Geovisualization*. CRC Press.

[5] Martins, N.C., Marques, B., Alves, J., Araújo, T., Dias, P. and Santos, B.S., 2022. Augmented reality situated visualization in decision-making. *Multimedia Tools and Applications*, 81(11), pp. 14749–14772.

[6] Chen, C., Liang, R., Pan, Y., Li, D., Zhao, Z., Guo, Y. and Zhang, Q., 2022. A Quick Development Toolkit for Augmented Reality Visualization (QDARV) of a Factory. *Applied Sciences*, 12(16), pp. 8338.

[7] Ayyanchira, A., Mahfoud, E., Wang, W. and Lu, A., 2022. Toward cross-platform immersive visualization for indoor navigation and collaboration with augmented reality. *Journal of Visualization*, 25(6), pp. 1249–1266.

[8] Ekanayake, I. and Gayanika, S., 2022, May. Data Visualization Using Augmented Reality for Education: A Systematic Review. In *2022 7th International Conference on Business and Industrial Research (ICBIR)* (pp. 533–537). IEEE.

[9] Mourtzis, D., Angelopoulos, J. and Panopoulos, N., 2022. Challenges and opportunities for integrating augmented reality and computational fluid dynamics modeling under the framework of industry 4.0. *Procedia CIRP*, 106, pp. 215–220.

[10] Sujihelen, L., Boddu, R., Murugaveni, S., Arnika, M., Haldorai, A., Reddy, P.C.S., Feng, S. and Qin, J., 2022. Node replication attack detection in distributed wireless sensor networks. *Wireless Communications and Mobile Computing*, 2022, pp. 1–11.

[11] Fombona-Pascual, A., Fombona, J. and Vicente, R., 2022. Augmented reality, A review of a way to represent and manipulate 3D chemical structures. *Journal of Chemical Information and Modeling*, 62(8), pp. 1863–1872.

[12] Rydell, L., 2022. Predictive algorithms, data visualization tools, and artificial neural networks in the retail meta verse. *Linguistic and Philosophical Investigations*, (21), pp. 25–40.

[13] Hirve, S.A. and CH, P.R., 2022. Improving big data analytics with interactive augmented reality. *International Journal of Information System Modeling and Design (IJISMD)*, 13(7), pp. 1–11.

[14] Lazaroiu, G., Androniceanu, A., Grecu, I., Grecu, G. and Neguriță, O., 2022. Artificial intelligence-based decision-making algorithms, Internet of Things sensing networks, and sustainable cyber-physical management systems in big data-driven cognitive manufacturing. *Oeconomia Copernicana*, 13(4), pp. 1047–1080.

[15] Roethe, A.L., Rösler, J., Misch, M., Vajkoczy, P. and Picht, T., 2022. Augmented reality visualization in brain lesions: A prospective randomized controlled evaluation of its potential and current limitations in navigated microneurosurgery. *Acta Neurochirurgica*, 164(1), pp. 3–14.

[16] Devagiri, J.S., Paheding, S., Niyaz, Q., Yang, X. and Smith, S., 2022. Augmented reality and artificial intelligence in industry: Trends, tools, and future challenges. *Expert Systems with Applications*, 207, pp. 118002.

[17] Liu, L., Shafiq, M., Sonawane, V.R., Murthy, M.Y.B., Reddy, P.C.S. and Kumar Reddy, K.C., 2022. Spectrum trading and sharing in unmanned aerial vehicles based on distributed blockchain consortium system. *Computers and Electrical Engineering*, 103, pp. 108255.

[18] Ashok, K., Boddu, R., Syed, S.A., Sonawane, V.R., Dabhade, R.G. and Reddy, P.C.S., 2023. GAN Base feedback analysis system for industrial IOT networks. *Automatika*, 64(2), pp. 259–267.

[19] Ashreetha, B., Devi, M.R., Kumar, U.P., Mani, M.K., Sahu, D.N. and Reddy, P.C.S., 2022. Soft optimization techniques for automatic liver cancer detection in abdominal liver images. *International Journal of Health Sciences*, 6, pp.10820–10831.

[20] Gholizadeh, M., Bakhshali, M.A., Mazlooman, S.R., Aliakbarian, M., Gholizadeh, F., Eslami, S. and Modrzejewski, A., 2022. Minimally invasive and invasive liver surgery based on augmented reality training: A review of the literature. *Journal of Robotic Surgery*, 17, pp. 1–11.

[21] Krüger, J.M., Palzer, K. and Bodemer, D., 2022. Learning with augmented reality: Impact of dimensionality and spatial abilities. *Computers and Education Open*, 3, pp. 100065.

[22] Singhal, A., Varshney, S., Mohanaprakash, T.A., Jayavadivel, R., Deepti, K., Reddy, P.C.S. and Mulat, M.B., 2022. Minimization of latency using multitask scheduling in industrial autonomous systems. *Wireless Communications and Mobile Computing*, 2022, pp. 1–10.

[23] Reddy, P.C.S., Yadala, S. and Goddumarri, S.N., 2022. Development of rainfall forecasting model using machine learning with singular spectrum analysis. *IIUM Engineering Journal*, 23(1), pp. 172–186.

[24] Angelopoulos, J. and Mourtzis, D., 2022. An intelligent product service system for adaptive maintenance of Engineered-to-Order manufacturing equipment assisted by augmented reality. *Applied Sciences*, 12(11), pp. 5349.

[25] Reddy, P.C.S., Suryanarayana, G. and Yadala, S., 2022, November. Data Analytics in Farming: Rice Price Prediction in Andhra Pradesh. In *2022 5th International Conference on Multimedia, Signal Processing and Communication Technologies (IMPACT)* (pp. 1–5). IEEE.

[26] Srilatha, D., and Thillaiarasu, N. 2023. "Implementation of Intrusion detection and prevention with deep learning in cloud computing." *Journal of Information Technology Management* 15.Special Issue: 1–18.

[27] Shyamambika, N., and Thillaiarasu, N. 2016. "A survey on acquiring integrity of shared data with effective user termination in the cloud." *2016 10th International Conference on Intelligent Systems and Control (ISCO)*. IEEE.

[28] Thillaiarasu, N., and ChenthurPandian, S. 2016. "Enforcing security and privacy over multi-cloud framework using assessment techniques." *2016 10th International Conference on Intelligent Systems and Control (ISCO)*. IEEE.

[29] Preethi, P., Asokan, R., Thillaiarasu, N., and Saravanan, T. 2021. An effective digit recognition model using enhanced convolutional neural network based chaotic grey wolf optimization. *Journal of Intelligent & Fuzzy Systems*, 41(2), 3727–3737.

[30] Thillaiarasu, N., Pandian, S. C., Vijayakumar, V., Prabaharan, S., Ravi, L., & Subramaniyaswamy, V. 2021. Designing a trivial information relaying scheme for assuring safety in mobile cloud computing environment. *Wireless Networks*, 27, 5477–5490.

[31] Shenoy, Ashwin, and Thillaiarasu, N. 2022. "A Survey on Different Computer Vision Based Human Activity Recognition for Surveillance Applications." *2022 6th International Conference on Computing Methodologies and Communication (ICCMC)*. IEEE.

[32] Gothai, E., Muthukumaran, V., Valarmathi, K., Sathishkumar, V. E., Thillaiarasu, N., and Karthikeyan, P. 2022. Map-reduce based distance weighted k-nearest neighbor machine learning algorithm for big data applications. *Scalable Computing: Practice and Experience*, 23(4), 129–145.

[33] Kaladevi, A. C., Saravanakumar, R., Veena, K., Muthukumaran, V., Thillaiarasu, N., and Kumar, S. S. 2022. Data analytics on eco-conditional factors affecting speech recognition rate of modern interaction systems. *Journal of Mobile Multimedia*, 18, 1153–1176.

[34] Srilatha, Doddi, and Thillaiarasu, N. 2022. "DDoSNet: A Deep Learning Model for detecting Network Attacks in Cloud Computing." *2022 4th International Conference on Inventive Research in Computing Applications (ICIRCA)*. IEEE.

[35] Srilatha, D., and Thillaiarasu, N. 2022, September. OIDCBMS: A Novel Neural Network based Intrusion Detection System to Cloud Computing based on New Cube Algorithm. *In 2022 4th International Conference on Inventive Research in Computing Applications (ICIRCA)* (pp. 1651–1656). IEEE.

[36] Reddy, P.C.S., Sucharitha, Y. and Narayana, G.S., 2021. Forecasting of Covid-19 virus spread using machine learning algorithm. *International Journal of Biology and Biomedicine*, 6, pp. 11–22.

[37] Reddy, P.C. and Sureshbabu, A., 2019. An adaptive model for forecasting seasonal rainfall using predictive analytics. *International Journal of Intelligent Engineering and Systems*, 12(5), pp. 22–32.

[38] Rupa, C., Srivastava, G., Ganji, B., Tatiparthi, S.P., Maddala, K., Koppu, S. and Chun-Wei Lin, J., 2022. Medicine drug name detection based object recognition using augmented reality. *Frontiers in Public Health*, 10, pp. 948.

[39] Morimoto, T., Kobayashi, T., Hirata, H., Otani, K., Sugimoto, M., Tsukamoto, M., Yoshihara, T., Ueno, M. and Mawatari, M., 2022. XR (extended reality: virtual reality, augmented reality, mixed reality) technology in spine medicine: status quo and quo vadis. *Journal of Clinical Medicine*, 11(2), pp. 470.

[40] Newbury, R., Cosgun, A., Crowley-Davis, T., Chan, W.P., Drummond, T. and Croft, E.A., 2022, August. Visualizing Robot Intent for Object Handovers with Augmented Reality. *In 2022 31st IEEE International Conference on Robot and Human Interactive Communication (RO-MAN)* (pp. 1264–1270). IEEE.

[41] Shaker Reddy, P.C. and Sureshbabu, A., 2020. An enhanced multiple linear regression model for seasonal rainfall prediction. *International Journal of Sensors Wireless Communications and Control*, 10(4), pp. 473–483.

[42] Reddy, P.C., Nachiyappan, S., Ramakrishna, V., Senthil, R. and Sajid Anwer, M.D., 2021. Hybrid model using scrum methodology for software development system. *Journal of Nuclear Energy Science & Power Generation Technology*, 10(9), pp. 2.

[43] Hasan, S.M., Lee, K., Moon, D., Kwon, S., Jinwoo, S. and Lee, S., 2022. Augmented reality and digital twin system for interaction with construction machinery. *Journal of Asian Architecture and Building Engineering*, 21(2), pp. 564–574.

[44] Rajaratnam, D., Weerasinghe, D.M.L.P., Abeynayake, M., Perera, B.A.K.S. and Ochoa, J.J., 2022. Potential use of augmented reality in pre-contract design communication in construction projects. *Intelligent Buildings International*, 14(6), pp. 661–678.

[45] Mihăilă, R. and Branişte, L., 2021. Digital semantics of beauty apps and filters: big data-driven facial retouching, aesthetic self-monitoring devices, and augmented reality-based body-enhancing technologies. *Journal of Research in Gender Studies*, 11(2), pp. 100–112.

[46] Siwach, G., Haridas, A. and Bunch, D., 2022, November. Inferencing Big Data with Artificial Intelligence & Machine Learning Models in Metaverse. *In 2022 International Conference on Smart Applications, Communications and Networking (SmartNets)* (pp. 01–06). IEEE.

Cloud and Security

6 Mathematical Model for Service-Selection Optimization and Scheduling in Cloud Manufacturing Using Sub-Task Scheduling with Fuzzy Inference Rule

*R. Murugesan, G.K. Jagatheswari,
and M. Rajeshwari*

6.1 INTRODUCTION

Industry 4.0 is the fourth industrial revolution, and Industrial Internet is the third wave that came after the first two waves, the Industrial Revolution and the Internet Revolution, according to Liu and Xu [1]. A service-oriented networked manufacturing paradigm is called cloud manufacturing. It is evident that various viewpoints were taken when these concepts of "advanced manufacturing models," "industrial revolution," and "industrial revolution plus Internet revolution" were promoted. Industry 4.0 and cloud manufacturing are both solely centered on manufacturing, but the latter takes manufacturing-specific challenges into greater consideration than the former [1,2]. The Industrial Internet, however, is applicable across numerous industries including the manufacturing industry. In this regard, the Industrial Internet concept has a wider scope than Industry 4.0, which in turn has a wider scope than cloud manufacturing. They all use emerging Internet, computer, and information technologies like IoT, CPS, cloud, and big data to fully realize the potential of industrial and manufacturing [1,2]. Cloud manufacturing is a network-based manufacturing model that offers data on production, distribution, description, and registration of exploited shared manufacturing resources [3]. One of the trends that aids in the success of the manufacturing sector is cloud manufacturing.

DOI: 10.1201/9781003388241-9

A new service-oriented networked manufacturing paradigm known as cloud manufacturing was introduced in 2010 to handle more challenging manufacturing challenges and to execute larger-scale collaborative manufacturing; it was influenced by cloud computing [4]. Since the start of cloud manufacturing research more than a decade ago, substantial progress has been made [5–7]. People are starting to understand how important cloud manufacturing is to the manufacturing industry, and both academic and industrial adoption of cloud manufacturing is growing rapidly [2,8, 9]. Researchers from various fields have shared their differing perspectives on cloud manufacturing. There has also been research into and testing of a preliminary use of cloud manufacturing in a number of sectors. They all contribute significantly to the development and use of cloud manufacturing [10–14]. There needs to be more discussion and arrangement reached on the correct understanding of the idea (its goal, scope, and boundary), the operation model, the service mode, the architecture, the essential characteristics, etc. [15–18]. The links between cloud manufacturing and some related concepts have rarely been emphasized. Furthermore, cloud manufacturing has many definitions, but none of them are presently standardized [19]. There aren't many books that discuss the factors, difficulties, and phases of cloud manufacturing implementation in great detail. Lack of consensus, in-depth discussions about the connections between cloud manufacturing and some related concepts, and a standardized definition are obstacles to the future growth of cloud manufacturing. At this crucial juncture when cloud manufacturing is going through tremendous development, it is essential to have in-depth discussions about the challenges and offer future views [20–23].

Cloud manufacturing enables productive communication between the enterprise cluster and the production chain. The diverse character of the manufacturing sector necessitates the use of a cloud-based intelligent scheduling strategy to manage production. Customers with no involvement in the design, engineering, or production process can purchase finished goods from conventional product-oriented manufacturing systems. However, in today's world where the customer is to decide and can do engineering on a product, customers require customization options in order to have access to goods that are suited to their particular needs. Small and medium-sized enterprises (SMMEs) have shown great potential in meeting customers' evolving expectations because of their wide range of functionalities, flexibility because they are smaller than large manufacturing enterprises, and their potential to grow as service-oriented businesses [1,2]. However, the absence of strong platforms that allow interactions between customers and SMMEs [24] and collaboration among geographically dispersed SMMEs in a real-time, on-demand, dynamic setting [25, 26] hinders the process of moving to a service-oriented manufacturing paradigm.

Cloud manufacturing offers an alluring solution for these problems by enabling the integration of the manufacturing resources of partner SMMEs in a way that allows them to carry out complex manufacturing tasks cooperatively with a high level of customization [19], a process that is not feasible in the conventional large-scale manufacturing systems. This platform is an on-demand search and suggestion tool that finds all kinds of manufacturing services such as design, engineering, machining, testing, and packing in the product lifecycle in order to meet the demands of customers for specialized manufacturing tasks. The best method to select and arrange these services is still a challenge, especially when factoring in transportation between widely dispersed SMME locations.

In this chapter, we wished to address the following: (1) How to plan and select services for manufacturing tasks in the cloud with different sub-task composition patterns. (2) The most efficient method to route a shipment between two geographically distinct SMMEs using a hybrid transportation network. (3) How to manage service usage and assign auxiliary duties to available services. One of the most popular combinatorial optimization issues is task management, scheduling the task while taking the limits into consideration is the key goal. There are limitations on the resources that are available and the time allotted, finding a suitable order in which to complete tasks is the goal. The majority of decision-makers in the actual world struggle with multi-objective scheduling issues. There is no single objective in this problem that can be used to assess the effectiveness of the proposed solution; rather, there are a number of objectives, some of which are time-based (e.g., minimizing the timeliness, tardiness, and due date) and others of which are cost-based (e.g., minimizing the cost of the manufacturing process, transportation costs, etc.). In these situations, there isn't a single solution that simultaneously meets all the goals, hence a compromise solution must be found in accordance with how the decision variable performs. Here the sub-task is decomposing a work into numerous smaller tasks, each of which is well suited to a single machine and has distinct execution requirements, and is the best approach to take advantage of the best solution.

Each sub-task is allocated to a machine and scheduled to be completed to minimize the completion time. This assignment problem is sometimes referred Non-Polynomial (NP) – hard problem [27]. To address this scheduling problem, a number of approximation and heuristic techniques have been presented [27–31]. These approaches are distinct from more traditional approaches like mathematical programming and the priority rule, as well as meta-heuristic and artificial intelligence-based algorithm approaches, including genetic algorithms, neural networks, particle swarm optimization, and the Artificial Immune System (AIS) algorithm [30,31].

The disadvantage of these approaches is that while heuristic algorithms can produce an optimal schedule quickly, they cannot ensure its quality, whereas approximation techniques can produce an ideal schedule quickly but require a considerable run time when dealing with big cases. The manufacturing industries are rapidly changing as a result of the growing need for personalization. Cooperation, adaptability, and agility are essential for manufacturing businesses if they want to thrive and compete. Consequently, manufacturing sectors are implementing contemporary technologies. Therefore, manufacturing businesses must improve their capabilities by becoming service providers in order to satisfy customer expectations.

The work presented in this chapter uses a mathematical model known as the Mixed Integer Linear Programming Model to examine the size of the problems that can be solved optimally and to create a sub-task scheduling method to address complex scheduling issues for customer orders. Additionally, attention was given to scheduling sub-tasks into available services, managing customer order occupancy, and allocating a logistic resource between small and medium-sized businesses. The numerical outcomes demonstrated that the suggested algorithms are capable of producing optimal results in a reasonable amount of computational time.

The remaining sections are organized as follows. Sections 6.2 and 6.3 contain the problem description for your convenience. Additionally, the following section demonstrates the building of MIP models. The SSOS problem formulation and mathematical models for service selection and scheduling for different sub-task composition structures are established in Section 6.3. Section 6.6 concludes this study and provides suggestions for future research, while Section 6.5 presents the computational results.

6.2   PROBLEM DESCRIPTION

In this chapter, we considering the customer order as a multi-task manufacturing service-oriented task. The given order can be done by the deadline with an average quality level of order at least MQ_{min}%, and this information can be gathered through the SMMES grading system. The order may be fulfilled within expected cost, and it must be delivered within the permitted time frame. To solve problems with multifunctional manufacturing job performance, a collection of composite services can be combined, and the multifunctional manufacturing work was broken down into several sub-tasks. In accordance with the functional specifications of each of these smaller tasks, qualified candidate computational modules (CMs) are sought out and gathered into the proper candidate CMs. Using this knowledge, select the better composite CMs for each sub-task from the pool of compound CMs that are accessible, and then every possible CMs for each multifunctional multitask is generated. Using QoS criteria, select the best candidate CMs from the list of possibilities. In compliance with QoS limitations, the candidate CMs are organized into multifunctional multitasks, and each group is integrated to complete each multifunctional multitask.

In this problem, T refers to the task. Let $T = (T_1, T_2, T_3, ..., T_I)$ called the number of task and denoted T_i where ($1 \leq i \leq I$) is the requesting candidate task. The Quality of Service (QoS) covers $Q(T_i) = (Q_{t1}, Q_{t2}, ..., Q_{tl})$ for each job, after which T_i can be broken down into a number of sub-tasks, i.e., $T_i = (st_j^1, st_j^2, ..., st_j^i)$, where j is the number of sub-tasks and we assume n number of registered SMMEs. Now the quality CMs for each sub-task st_j^i are pooled into a candidate set of CMs: $MCs = \{MCs_1^j, MCs_2^j, ..., MCs_{kj}^j\}$. The numbers of CMs are related to each sub-task $(st_j^1, st_j^2, ..., st_j^j)$ in the workflow of a multi-functional oriented multi-task T_i.

6.3   SERVICE SELECTION

There will be functionally equivalent CMs with a different quality service metric available for each sub-task. Selecting SMMEs is challenging because, despite using the same resources, different SMMEs may have varying standards for completing the sub-task in terms of number of quality service indicators. To differentiate between the CMs candidates, we consider MC (Manufacturing cost), MT (Manufacturing time), and quality-metrics, are the main variables to be maximized in the service selection. The average pass rate is additionally used to determine the level of quality. In order to create the optimum cloud service selection, the outcome of aggregating the subsequent metric of all selected CMs should be a quality service measure.

Dealing with computer, manufacturing, and mixed-type sub-tasks are some of the more complex and difficult ones. In CMs, a mixed-structure sub-task can be transformed into a sequential one. Because it directly affects the CMs environment, study on logistics among SMMEs is essential. For the sub-task composition structure in this study, we recommended MIPM. The system is given a limited number of sub-tasks, and it must select the one that is the best fit. Any one of the accessible systems in the manufacturing structure must be processed by each sub-task. The probability that the system will select the st^{th} is $P_{prop} = 1$. Figure 6.1 below   shows the service selection, and Table 6.1 gives the notation that was employed to generate the SSOSP.

The Cost function in the selective sub-task selection among the alternative sub-tasks is $Min \ C_{min} = \sum_{st=1}^{p} P_{prob} CSST_{st}$

And the Task completion time in the selective sub-task selection among the alternative sub-tasks is

$$Min \ T_{min} = \sum_{st=1}^{p} P_{prob} TSST_{st}$$

In this chapter, the major objective is to cut back on overall costs, production time, and transportation time. In this problem, the manufacturer used to estimate the cost based on the quotation that the organization provided for each sub-task, but for a variety of reasons, certain prices were overlooked when compared to the real costs. Here, we consider a number of factors, including the weight and path of the items as well as the price and length of the journey. These components provide the framework for constructing the equation (6.1):

$$C \ Tr_{i,j} = D_{i,j} CTr_{unit} \tag{6.1}$$

$$Tr \ T_{i,j} = D_{i,j} TrT_{unit} \tag{6.2}$$

The variable cost for delay of delivery on shipments from individually destination is also defined in:

$$Min \ C_{min} = \sum_{st=1}^{p} \sum_{r \in R_{st}} C_{Mfg} x_{st,r} + \sum_{q=1}^{p} \sum_{i=1}^{m} \sum_{j=1}^{n} CTr_{i,j} w_{i,j} y_{q,q+1}^{i,j} P_{prob} \tag{6.3}$$

Equation (6.3) states that the project completion time should be kept to the minimum. This completion time includes manufacturing time, transport time among destinations, and resources waiting time (6.4). A manufacturer commits to a specific period of time, which includes setup, processing, and maintenance time, when they send a manufacturing sub-task to a manufacturing facility. Equation (6.2) states that the unit distance between the centers and the time spent for the waiting period

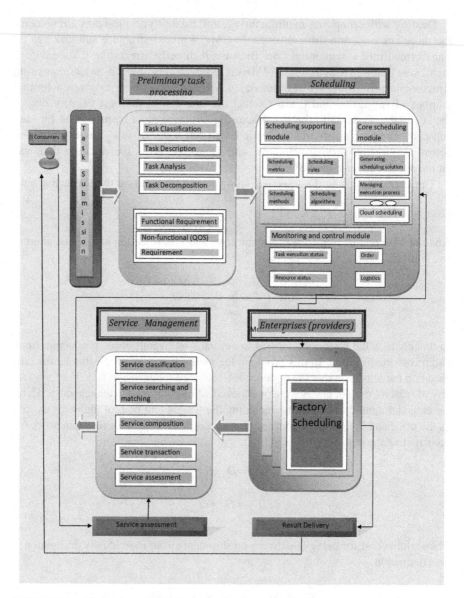

FIGURE 6.1   Illustrations of the services being selected.

before the production process starts affect the length of the time period needed for transportation.

$$Min \ T_{min} = \sum_{st=1}^{p} \sum_{r \in R_{st}} T_{Mfg} x_{st,r} + \sum_{q=1}^{p} \sum_{i=1}^{m} \sum_{j=1}^{n} TrT_{i,j} w_{i,j} y_{q,q+1}^{i,j} P_{prob} \qquad (6.4)$$

TABLE 6.1
Notation for SSOSP

CSST	cost in the selective sub-task	Tr	Transportation
TSST	completion time of selective sub-task	TrTij	Transportation time between the SMMEs
I	Multiple Tasks	$X_{st,r}$	1 if st^{th} sub-task is performed using r^{th} MCs's
1	Index of CMs	$Ships_{tart,st}$	Shipment from the starting point of supplier of first sub-task
I,j	Index for location of SMMEs	P	Index of manufacturing sub-tasks (st,u=1,...,st)
St, u	Index for sub-task	LOC_{end}	Location of transportation ends
CTr	Cost of transportation	LOC_{start}	Location of transportation starts
TrT	Time of transportation	$y^{I,j}_{start}$	1 if i^{th} and p^{th} sub-task performed location I and j respectively and there is a direct transportation between these locations
			0 otherwise
C_{mgf}	Manufacturing cost	$MgfT_{st}$	Manufacturing time for sub-task
$X_{I,j}$	1 if st^{th} sub-task is performed using r^{th} CMs; 0 otherwise	SMgf(Pu)	Starting time of manufacturing process st^{th} sub-task.
R `	Index of SMMEs	Emgf(Pu)	End time of manufacturing process st^{th} sub-task
R_{st}	Pool of cloud manufacturing qualities services set for st^{th} sub-task	$\mu_{u,k}$	1 if st^{th} sub-task is performed using r^{th} CMs before starting its occupied time; 0 otherwise
q	Index of supplying and delivering sub-task in addition to manufacturing sub-task (q,p=start,1,...,st,...,end)	C	Cost
$w_{i,j}$	Transportation weight from between q and q+1 sub-task	T	Time
$y^{i,j}_{q,q+1}$	1 if transportation from I to j,q^{th} sub-task performed in I,j	MC_{cus}	Maximum cost that customer willing to pay
T_{min}	Time minimization	DDc_{us}	Production delivery deadline specified by the customer
C_{min}	Cost minimization	Tr_{unit}	Transportation per unit
T_{mgf}	Manufacturing time	TrT_{unit}	Transportation time per unit
D	Distance		

The mean pass rate of the chosen SMMEs is used to calculate the other target function, known as quality of service metrics (maximum quality), in the following equation.

$$MQ_{min} = \frac{1}{s}\sum_{s=1}^{s}\sum_{st \in R_{st}} pass_r x_{st,r} \qquad (6.5)$$

We used the various limits indicated below in order to maximize the primary objective function at the same time. Constraint (6.6), which stipulates that a shipment must be made from the supplier's starting point to the position of the first sub-task, provides the mathematical formulation. Equation (6.7) is the constraint to provide transportation between the production plants to perform the next sub-tasks. The last manufacturing center's delivery to the client is represented by constraints (6.8) and (6.9) that demand that only one manufacturing center complete a sub-task.

$$x_{st,r} \le ship_{start,st}^{Loc_{start}Loc_r} \forall r \in R_{st} \qquad (6.6)$$

$$x_{st,r} + x_{st+1,k} \le 1 + ship_{1,st}^{Loc_1,Loc_r} \forall st < P, r \in R_{st}, k \in R_{st+1} x_{st,r} \le ship_{start,st}^{Loc_{start}Loc_r} \forall r \in R_{st} \quad (6.7)$$

$$x_{st,r} \le ship_{start,st}^{Loc_{start}Loc_r} \forall r \in R_{st} \qquad (6.8)$$

$$\sum_{r \in R_{st}} x_{st,r} = 1 \forall st \qquad (6.9)$$

Equations (6.10) and (6.11) provide estimates for the beginning and end of each sub-task, respectively, because the start of one sub-task depends on the finish of the sub-task that comes before it. In rare circumstances, there may also be a need for waiting and transit time between two linked sub-tasks. The starting time of the sub-task should be calculated as the sum of the manufacturing time, transportation time, all preceding sub-tasks, and their waiting times. The total of a sub-task's starting time and associated manufacturing time determines how long it will take to complete.

$$\sum_{i=1}^{m}\sum_{j=1}^{n} TrT_{i,j} y_{start,1}^{i,j} + \sum_{st=1}^{p-1}\sum_{r \in R_{st}} MfgT_{st,r} x_{st,r} + \sum_{q=1}^{p-1}\sum_{i=1}^{m}\sum_{j=1}^{n} TrT_{i,j} y_{q,q+1}^{i,j} \quad \forall P \qquad (6.10)$$

$$SMfgP_u + \sum_{k \in R_u} MfgT_{u,k} x_{u,k} = EMfgP_u \quad \forall u \qquad (6.11)$$

The equations (6.12) and (6.13) define the limitation of the constraints that are utilized to plan a sub-task to specific cloud manufacturing centers from beginning to end before occupancy. For each sub-task, this variable allows only one of these constraints to be in effect. The constraints (6.12) and (6.13) provide the specified quality of service requirements of the customer in terms of cost and time, and the last constraint (6.16) places limitation on the pertinent decision variables.

$$SMfgP_u + Mfg \quad \mu_{u,k} \geq EMfg_o Tx_{u,k} \quad \forall u, k \in R_u \tag{6.12}$$

$$EMfgP_u \leq SMfg_0 Tx_{u,k} + Mfg(1 - \mu_{u,k}) \quad \forall u, k \in R_u \tag{6.13}$$

$$C \leq MC_{cus} \tag{6.14}$$

$$T \leq DD_{cus} \tag{6.15}$$

$$x_{st,r}, \mu_{u,k}, y_{i,j}, y_q^{i,j} \in [0,1]$$
$$SMfgP_u, EMfgP_u, w_{i,j} \geq 0, \forall i, j, st, r, k, u, p \tag{6.16}$$

6.4   OVERALL OBJECTIVE FUNCTION

Maximizing manufacturing quality for the customer order while reducing cost and time is the recommended strategy for dealing with SSOSP. There is no optimal approach that can achieve all three objectives. We were able to optimize the objective function overall by combining the three objectives. The measurement and the range of the goal function may be very different. As a result, it is not possible to simply sum the several objective functions to produce the overall objective function. We provided our fuzzy logic approach to address this issue in order to determine the optimal value of overall goal of the function.

In order to cope with issues where there are sources of ambiguity, Zadeh introduced fuzzy logic. Instead of using the binary values 0 and 1, fuzzy logic offers multi-valued membership functions to create an objective function. Let x be the element of the fuzzy set X of objectives function and the membership in fuzzy set A of X is the characteristic function of μ_A such that $\mu_A = \begin{cases} 1 & \textit{iff } x \in A \\ 0 & \textit{otherwise} \end{cases}$.

Input and output spaces are defined as maps for the fuzzy inferences. Rules of inference, sometimes known as "IF-THEN" statements, are written as $IF(antecedent)THEN(consequents)$. Fuzzification is the act of converting any input data into fuzzy membership values, while defuzzification is the process of getting a crisp value from the inferences. We must recognize the fuzzy variable, which has uncertain values or ambiguous limits, in order to determine the objective function's ideal value. The fuzzy variables used in this study are MC, MT, and MQ, which stand for manufacturing cost, manufacturing time, and manufacturing quality. Greater importance placed on quality and lesser value placed on price and turnaround time are in the customer's best service; as a result, CMs must be chosen based on customer interest.

$MC = (current\ value - minimal\ values)/(maximal\ values - minimal\ values)$ the similar computations were also used for MT (Mean Time), and MQ (Mean Quality). These factors provide useful data for determining the CMs based on client interest. The variables are summarized as follows

$$MC = \begin{cases} high\left(\mu_{mch}\right) \\ average\left(\mu_{mca}\right), \\ low\left(\mu_{mcl}\right) \end{cases} MT = \begin{cases} high\left(\mu_{mth}\right) \\ average\left(\mu_{mta}\right), \\ low\left(\mu_{mtl}\right) \end{cases} MQ = \begin{cases} high\left(\mu_{mqh}\right) \\ average\left(\mu_{mqa}\right) \\ low\left(\mu_{mql}\right) \end{cases}$$

The membership function, which is linguistically labeled for specifying the domain of discourse of fuzzy variable, must then be defined.

The customer concern is influenced by more than just price and quality; delivery performance is also very important. As a result, production time is the third priority, cost is the second, and the fuzzy variable we consider manufacturing quality is the first.

The fuzzy min-max composition method is

$$C = A \circ B \in X \times Y \rightarrow [0,1] \, and \, \mu_C\left(x,y\right) =$$
$$\sup_{z \in \mu z} min\{\mu_A\left(x,z\right), \mu_B\left(z,y\right) \, where \, A = f\left(X,Z\right),$$
$$B = f\left(Z,Y\right), \forall\left(x,y\right) \in X \times Y$$

$$If \, \forall x \in U, \, then \, A \cup B = \{\mu_x\left(x\right) = max\left(\mu_A\left(x\right), \mu_B\left(x\right)\right)$$

and the overall objective function is defined as

$$O_{obj} = max\{min(\mu_{mch}, \mu_{mth}, \mu_{mqh}), min(\mu_{mch}, \mu_{mth}, \mu_{mqa}), min(\mu_{mch}, \mu_{mth}, \mu_{mql}),$$
$$min(\mu_{mch}, \mu_{mta}, \mu_{mqh}), min(\mu_{mch}, \mu_{mta}, \mu_{mqa}), min(\mu_{mch}, \mu_{mta}, \mu_{mql}),$$
$$min(\mu_{mch}, \mu_{mtl}, \mu_{mqh}), min(\mu_{mch}, \mu_{mtl}, \mu_{mqa}), min(\mu_{mch}, \mu_{mtl}, \mu_{mql}),$$
$$min(\mu_{mca}, \mu_{mth}, \mu_{mqh}), min(\mu_{mca}, \mu_{mth}, \mu_{mqa}) min(\mu_{mca}, \mu_{mth}, \mu_{mql}),$$
$$min(\mu_{mca}, \mu_{mta}, \mu_{mqh}), min(\mu_{mca}, \mu_{mta}, \mu_{mqa}), min(\mu_{mca}, \mu_{mta}, \mu_{mql}),$$
$$min(\mu_{mca}, \mu_{mtl}, \mu_{mqh}), min(\mu_{mca}, \mu_{mtl}, \mu_{mqa}), min(\mu_{mca}, \mu_{mtl}, \mu_{mql})\}$$

6.5   RESULT AND DISCUSSION

We believed that a CM is providing 10 SMMEs with a platform for a system of service-oriented manufacturing. A directed network can be used to route the logistics between the various SMMEs. A consumer placing an order and asking for a bespoke order with minimal cost, a set deadline, and average quality. The values of customer-promised cost, customer-delivered time, customer-expected quality, and CTr unit are shown in Table 6.4 along with other parameters used in the SSOSP. The CTr time per unit parameter was selected based on real-time settings, while the other parameters were determined at random. Before transportation is completed, the task designated as a sub-task must be completed. Transportation begins at the supplier's site. In this work we defined the distance d_{ij} between the enterprises, as given in Table 6.2.

TABLE 6.2
Distance between the Enterprises

SMMEs Index	1	2	5	6	7	9	11	15	16	17
1	0	242	332	432	128	365	463	275	244	360
2	242	0	126	223	435	342	321	248	144	343
5	332	126	0	410	330	394	475	371	215	199
6	432	223	410	0	183	184	187	425	214	215
7	128	435	330	183	0	299	135	340	434	234
9	365	342	394	184	299	0	174	247	496	343
11	463	321	475	187	135	174	0	125	217	187
15	275	248	371	425	340	247	125	0	240	187
16	244	144	215	214	434	496	217	240	0	197
17	360	343	199	215	234	343	187	213	197	0

TABLE 6.3
Sub-task Cost, Time, and Quality with Respect to SMMEs

S.No Sub-task	SMMEs Index	Pass rate	Occupied time	Manufacturing cost per unit (Rs in ten thousand)/ Time(week) per unit					
				st1	st2	st3	st4	st5	st6
1	1	0.93	2-4	3.5/2			3.2/2.5		
2	2	0.75	0-3		8.3/2.1				
3	5	0.70	3-4	9.2/1.2					
4	6	0.95	2-7		10.2/2.5			10.7/2.3	
5	7	0.87	2-4			13.2/3			
6	9	0.76	5-7		8.2/1.2		9.3/2		
7	11	0.87	3-8	7.7/3				8.2/3	
8	15	0.82	2-6			11.2/3.5			12.1/4
9	16	0.81	4-7		4.2/2.3				
10	17	0.79	4-8	14.2/3					13.7/2.8

Additionally, the COSP is divided into six sub-tasks $st_k\, k = 1,2,\ldots6\,(st1, st2, st3, st4, st5, st6)$ and the CM platform pairs the services with the sub-tasks in accordance with their needs. A number of SMMEs, each with a different set of quality measurements, are available for each sub-task, and occasionally more than one SMME will use the same metrics. For the case of SMMEs and their cloud manufacturing indices, the generated statistics of manufacturing cost per unit and manufacturing time are presented in Table 6.3.

When using cloud manufacturing, the CMs platform connects services with sub-tasks according to their specifications and the user's consent, and then sends invitations to manufacture to the suitable services. Finding the best CMs to handle

TABLE 6.4
Parameters in Solving the SSOSP

S.No	Parameters	Values	S.No	parameters	Values
1	i: Location of SMMEs	17	6	w_t: time parameter	0.4
2	r: Index of MCs	20	8	w_q: quality parameter	0.2
3	st: Number of sub-task	6	9	C_M: max.cost customer willing to pay	44000
4	$W_{i,j}$: Weight of transportation	40 if i<6 and 200 if i>=7	10	U_{rc}: Unit transportation cost	0.003
5	Loc_{start}	15	11	U_{rr}: Unit time transportation cost	463×10^{-6}
6	Loc_{end}	10	12	T_{max}: deadline	14
7	w_c: cost parameter	0.4		Q_{max}:	75

each sub-task and the best SMMEs for routing are the goals of this effort, which is to find the optimal service selection and scheduling.

Transportation starts with supplier locations, as shown in Table 6.4. Six sub-tasks must be completed by 17 distinct SMMEs before transportation is complete. Additionally, Table 6.2 provides information on SMMEs, their CMs, and the travel time between the locations. We can determine the other parameters, such as $TC_{i,j}$ and $TT_{i,j}$, from Table 6.4. The resultant payout table's range of optimal solution values and the value of the overall Objective Function (OF) are shown in the table below.

When $W_C = 1$ and $W_T = W_Q = 0$, the lower bound value of the manufacturing cost of OF is equal to Rs 38.3 in thousands and the upper bound is equivalent to Rs 72.4 in thousands. The value of the objective function's quality is in the range of 81% to 95%, while the value of the time of the objective function is between 1.2 and 7.5 weeks. The total OF value, calculated using the min-max algorithm as specified by the min-max formula, is 0.73, the highest value. If transportation is excluded, the cost of the OF using the min-max algorithm is 53 thousand dollars, with a maximum time of 4 weeks and an average quality of 85.8%.

The optimal flow of the sub-tasks carried out by SMMEs is $11 \rightarrow 9 \rightarrow 15 \rightarrow 1 \rightarrow 6 \rightarrow 15$, with a unit transportation cost of Rs. 4.659 thousand. Here, two hypotheses are offered to account for these results. Without considering transportation costs, the first scenario is evaluated. The outcome indicates that the value of OF has been significantly decreased. The cost and duration of the second scenario are taken into account. Based solely on CMs, the model chooses SMMEs.

In practice, the constraints of customer usually include cost, delivery time, and quality. There is a hypothesis that states that rapid production delivery leads to more cost, or decreases the quality. An increase in C_{max} causes the higher cost of OF with time, and also it improves the value of the other two OFs. The quality values increase from 75% to 85%, during the time OF remains constant. The quality objective

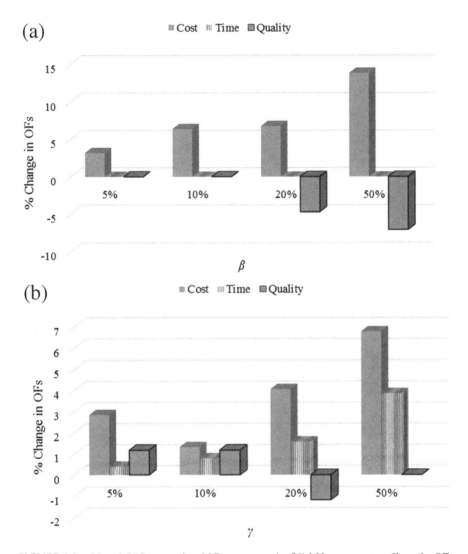

FIGURE 6.2   (a) and (b) Increase in a MCs parameter by β% Mfg$_{st}$ parameter γ% on the OFs.

function OF increase of 1% can be found that costs an almost 3% increase in the cost OF. Figure 6.2 (a) and (b) shows that the result of an increase in a MCs parameter by β% and Mfg$_{st}$ parameter γ% on the OFs.

In addition, research on the effects of water has shown that higher water levels have a negative effect on the value of all three OFs; better solutions for quality OF have a lower percentage of variance in their range values than the other two OFs. The following figure illustrates the impact of w$_c$; increasing the cost function causes a decrease in the time function. That is, if rising costs improves both cost and quality as shown in Figure 6.3 and Figure 6.4.

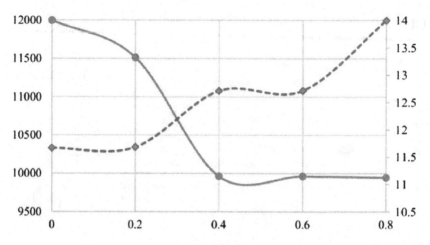

FIGURE 6.3   Increasing w_c on the value of cost and time OFs.

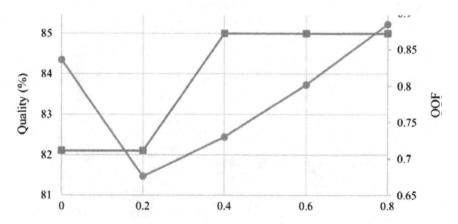

FIGURE 6.4   The value of quality OF and OOF.

6.6   CONCLUSION

In this research, we looked at manufacturing costs, production times, transportation costs, and transit times. For CM, a multitask scheduling paradigm based on fuzzy rules is suggested. Here, we took into account production costs, production times, transit times between small- and medium-sized manufacturing businesses, and quality as the constraints. All three objective functions were identified as small, medium, or high based on the proposed fuzzy rule, and the best order for the tasks to be completed was determined using the min-max rule. The primary important aspect in cloud manufacturing is transportation because it has a considerable impact on work scheduling. The sub-tasks and service utilization were met via dynamic calculation. The outcomes of our min-max algorithm showed that scheduling heavier workload activities with a high priority has a better chance of achieving a speedy optimal result. Future work will be taken into consideration for continual customer order arrivals with various time constraints.

REFERENCES

1. Thillaiarasu N, Pandian SC, Vijayakumar V, Prabaharan S, Ravi L, Subramaniya swamy V (2021). 'Designing a trivial information relaying scheme for assuring safety in mobile cloud computing environment', *Wirel Netw* 27, 5477–5490.
2. Shenoy, Ashwin, Thillaiarasu N (2022). 'A Survey on Different Computer Vision Based Human Activity Recognition for Surveillance Applications,' *2022 6th International Conference on Computing Methodologies and Communication (ICCMC)*. IEEE.
3. Akbaripour H, Masehian E, Roostaei A (2017) 'Landscape analysis and scatter search metaheuristic for solving the incapacitated single allocation hub location problem', *Int J Ind Syst Eng*. 26(4):425–459.
4. Li W et al (2015) 'Subtask scheduling for distributed robots in cloud manufacturing', *IEEE Syst J*. 11(2):1–10.
5. Akbaripour H, Houshmand M, Valilai OF (2015) 'Cloud-based global supply chain: A conceptual model and multilayer architecture', *J Manuf Sci Eng*. 137(4):31–36.
6. Arora N et al (2008) 'Putting one-to-one marketing to work: Personalization, customization, and choice', *Mark Lett*. 19(3–4):305.
7. Aykin T (1995) 'The hub location and routing problem', *Eur J Oper Res*. 83(1):200–219.
8. Benayoun R et al (1971) 'Linear programming with multiple objective functions: Step method (STEM)', *Math Program* 1(1):366–375.
9. Campbell JF (1994) 'Integer programming formulations of discrete hub location problems', *Eur J Oper Res*. 72(2):387–405.
10. Cheng Y et al (2017) 'Modeling of manufacturing service supply– demand matching hyper network in service-oriented manufacturing systems', *Robot Comput Integr Manuf*. 45:59–72.
11. Cheng Z, et al (2014) 'Multitask oriented virtual resource integration and optimal scheduling in cloud manufacturing', *J Appl Math*. 2014(1):1–9.
12. Deb K, Miettinen K, Sharma D (2009) 'A hybrid integrated multiobjective optimization procedure for estimating nadir point', In *International Conference on Evolutionary Multi-Criterion Optimization*. Berlin, Heidelberg: Springer, pp. 569–583.
13. Huang B, Li C, Tao F (2013b) 'A chaos control optimal algorithm for QoS-based service composition selection in cloud manufacturing system', *Enterp Inf Syst*. 8(4):445–463.
14. Kurdi H et al (2015) 'A combinatorial optimization algorithm for multiple cloud service composition', *Comput Electr Eng*. 42: 107–113.
15. Lartigau J et al (2015) 'Cloud manufacturing service composition based on QoS with geo-perspective transportation using an improved Artificial Bee Colony optimisation algorithm', *Int J Prod Res*. 53(14):4380–4404.
16. Kesen SE, Das SK, Güngör Z (2010) 'A genetic algorithm based heuristic for scheduling of virtual manufacturing cells (VMCs)', *Comput Oper Res*. 37(6):1148–1156.
17. Srilatha Doddi, Thillaiarasu N (2023) 'Implementation of intrusion detection and prevention with deep learning in cloud computing.' *J. Inf. Technol. Manag*. 15.Special Issue: 1–18.
18. Thillaiarasu N, ChenthurPandian S (2016) 'Enforcing security and privacy over multi-cloud framework using assessment techniques.' *2016 10th International Conference on Intelligent Systems and Control (ISCO)*. IEEE.

19. Preethi P, Asokan R, Thillaiarasu N, Saravanan T (2021) 'An effective digit recognition model using enhanced convolutional neural network based chaotic grey wolf optimization,' *Journal of Intelligent & Fuzzy Systems* 41(2), 3727–3737.
20. Gothai E, Muthukumaran V, Valarmathi K, Sathishkumar VE, Thillaiarasu N, Karthikeyan P (2022) 'Map-reduce based distance weighted k-nearest neighbor machine learning algorithm for big data applications,' *Scalable Comput Pract Exp.* 23(4), 129–145.
21. Kaladevi AC, Saravanakumar R, Veena K, Muthukumaran V, Thillaiarasu N, Kumar SS (2022) 'Data Analytics on Eco-Conditional Factors Affecting Speech Recognition Rate of Modern Interaction Systems,' *J. Mob. Multimed.* 1153–1176.
22. Srilatha, Doddi, Thillaiarasu, N. (2022) 'DDoSNet: A Deep Learning Model for detecting Network Attacks in Cloud Computing,' *2022 4th International Conference on Inventive Research in Computing Applications (ICIRCA).* IEEE.
23. Song T et al (2014) 'Common engines of cloud manufacturing service platform for SMES', *Int J Adv Manuf Technol.* 73(1–4):557–569.
24. He W, Xu L (2014) 'A state-of-the-art survey of cloud manufacturing', *Int J Comput Integr Manuf.* 28(3):239–250.
25. Huang B, Li C, Yin C et al (2013a) 'Cloud manufacturing service platform for small- and medium-sized enterprises', *Int J Adv Manuf Technol.* 65(9–12):1261–1272.
26. Shyamambika N, Thillaiarasu N (2016) 'A survey on acquiring integrity of shared data with effective user termination in the cloud.' *2016 10th International Conference on Intelligent Systems and Control (ISCO).* IEEE.
27. Akbaripour H, Masehian E (2013) 'Efficient and robust parameter tuning for heuristic algorithms', *Int J Ind Eng Prod Res.* 24(2):143–150.
28. Banaszak ZA, Zaremba MB (2006) 'Project-driven planning and scheduling support for virtual manufacturing', *J Intell Manuf.* 17(6):641–651.
29. Bennett DP, Yano CA (2004) 'A decomposition approach for an equipment selection and multiple product routing problem incorporating environmental factors', *Eur J Oper Res.* 156(3):643–664.
30. Shen Y, Yang X (2011) 'A self-optimizing QoS-aware service composition approach in a context sensitive environment', *J Zhejiang Univ Sci C.* 12(3):221–238.
31. Silva APD, Stam A (1997) 'A mixed integer programming algorithm for minimizing the training sample misclassification cost in two-group classification', *Ann Oper Res.* 74:129–157.

IoT

7 Social Media Initiatives through IoT to Link the Bridge between Industrial Demands with Higher Education Millennial Students through Experience Learning

Dara Vijaya Lakshmi, M. Saradha, and Chitra Kesavan

7.1 INTRODUCTION

The availability of skilled labor is essential for the successful development of an industry. Industries have certain qualifications and skill requirements for people who are devoted to a career within their ranks. One of the main drivers of economic, cultural, and social progress is higher education. Yet, it has been noticed that recent Indian university graduates lack the credentials necessary to function well in fast-paced businesses. They also don't comprehend the demands and reality of their field. The business community has made it clear that students need to acquire work experience. In order to be successful in their jobs, students essentially need the right communication abilities to work across disciplinary and institutional barriers.

The level of higher education directly affects the industrial growth and, consequently, the socioeconomic development of the nation. The labor market demands that the workforce be up-to-date on the most recent innovations, but academia is having difficulty filling the gap. The disparity between the skills needed and those that are now accessible has only widened, aggravating India's already serious unemployment issue.

Higher education institutions must put a greater emphasis on addressing labor market demands and providing a better skilled workforce in order to compete internationally. Schools must also continue to play the traditional function of preparing upcoming "knowledge workers" for success in a society that is both complicated and increasingly competitive.

Although being highly valued by society, teachers in India get inadequate pay. Education is one of the most sought-after majors at Indian colleges, despite the lack of adequate compensation. Education is viewed as a motivator for advancement.

Due to lack of resources, poor infrastructure, low funding, and less professional development opportunities for professors, results in out-of-date abilities, and few or outdated teaching tools and materials. For example, the current low level of library use is a result of poorly maintained buildings and equipment at public universities as well as a large percentage of outdated content. Perhaps more crucially, Indian higher education has significant difficulties with the standard and applicability of its career preparation curricula.

7.2   RESEARCH GAP

There are many elements that influence the divide between academia and industry

1. The drawbacks of the existing higher education system, which include the following:
    a. Absence of practical knowledge among recent grads.
    b. Generic education that is too broad and doesn't put enough emphasis on specialization.
    c. A lack of communication between higher education institutions and stakeholders.
2. The academic staff's workload takes up the majority of their time and energy, making it difficult for them to interact with business experts. Teaching, participation in research and development, oversight of students' research projects, graduation projects, etc. are all part of the instructors' varied workload. Also, lecturers are in charge of creating new instructional materials and curricula. Teachers too require training and knowledge updates just as frequently and as much as students do, especially in today's instantaneous world.

7.3   OBJECTIVE

a. To analyze the relationship between different methods and systems of learning systems for implementing an effective learning process.
b. To understand project-based learning within a semantically integrated social network.
c. To estimate online education, specifically an e-learning system.

7.4   REVIEW OF LITERATURE

According to researchers, social media is used more frequently than learning management systems (LMS) provided by universities (Deng & Tavares, 2013; Jong et al., 2014; Bateman & Willems, 2012). Selwyn (2009) examined whether web-based learning is effective for higher education in the 21st century. He believed it to be a place where students might successfully negotiate a potential "role conflict" they might have in their interactions with faculty members, academic expectations, and university work.

Despite most students view social networks rather than a formal educational instrument (Selwyn, 2009). Stone and Logan (2018) studied the potential of social media platforms such as WhatsApp, Twitter, Facebook, and Instagram to promote active learning among students. It is arguable that the idea that learning is fundamentally a social activity emerged in the age of enlightenment and was heavily emphasized in the work of constructivists in the 19th century rather than with the usage of social media in educational situations (Crotty, 1998; Vygostsky, 1986). This study has important applications on conducting virtual activities for enhancing the motivation and encouragement of student-to-student interaction. Additionally, there was a wider variety of learning tools, less focus on direct instruction, and more emphasis on social networking and self-learning (Manca & Ranieri, 2013). The HEI, however, was expected to take the lead in facilitating discussions in small groups and encouraging students' independent study. And, like in any university course, the HEI was in charge of facilitating "graduate characteristics," or abilities considered essential for graduates by the architecture profession (Halliday, 1993). Recently, Henderson, Selwyn, and Aston (2017) discovered that, in addition to allowing for improved course structure, time savings, and geographical flexibility, digital technology had the key benefit of facilitating collaboration and communication among students. Students were thought to frequent and use social media the most regularly for talks (Junco, 2013; Kent and Taylor, 2014). The HEI anticipated that using it in the course would build a sense of community among peers and facilitate regular and consistent communication. Because students were likely already avid Facebook users (Bateman & Willems, 2012), the HEI saw it as a platform that would engage students and encourage their involvement. The CC's assumptions that such an approach would assist students are compatible with sociocultural learning theory, which highlights that learning is a social enterprise (Vygotsky, 1978). In order to evaluate the course redesign, the research team concluded that a model for interpreting the student learning experience in the course should serve as the study's framework. The CC opted to apply the researchers' proposed model for improving the student learning experience (Awidi, 2006). This model (MISL), which theorizes the components of the student learning experience (SLE) in a digital learning environment, falls within the constructivist paradigm. Later in the publication, this model is discussed along with how it was used in this investigation.

7.5 THEORETICAL FRAMEWORK

To store and retrieve enormous student learning datasets, a knowledge acquisition system and artificial cognitive declarative memory model are used. An online group learning course's performance data and student communications are monitored and aggregated using a Deep Academic Learning Intelligence system for machine learning-based student services. The system presents a series of suggestions based on communication activity, social activity, and academic success data, and then leverages feedback from answers and post-recommendation data to improve the machine learning-based system.

7.6 SYSTEM DIAGRAM

An e-learning method comprises the steps of:

a. Enabling a teacher to connect to a variety of users via the Internet using one of a number of remote data processing devices.
b. Enabling the teacher and users to use one of the remote data processing devices to access an online teaching database that connects a curriculum server, an examination server, a user records server, and an academic server over the Internet.
c. Permitting the instructor to use one of the remote data processing devices to access the curriculum server.
d. A process that includes the following steps for managing participant interaction in a social network using computers with nodes and storage.
e. A social network with linked computers that uses at least one of them to gather data for at least two users on the network and assigns each user to a computer node.
f. Finding each computer node's initial state using at least one other computer.
g. Finding individuals who are in comparable situations by locating similar computer nodes in the social network using at least one other computer.
h. Managing communication between participants in comparable situations by controlling the computer node each participant is assigned over at least one network as provided in Figure 7.1.

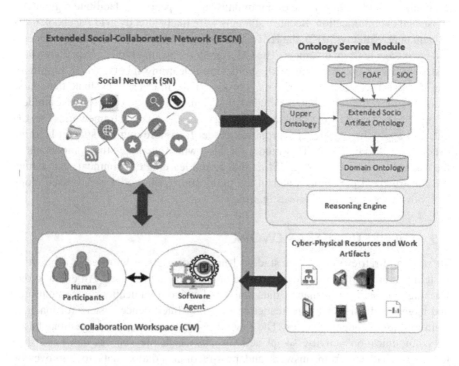

FIGURE 7.1   System design of social media adoption.

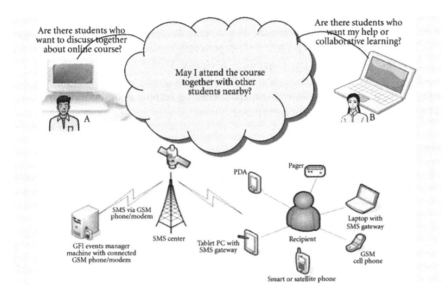

FIGURE 7.2 IoT driven livestreaming social media for interfaces.

i. The e-learning method is an online learning method, a distant learning method, or a classroom teaching method.

Traditional learning management systems (often referred to as "LMSs") offer an integrated system with facilities for organizing and managing online courses. An LMS typically manages enrollment, management of assignments, lesson plans, syllabuses, discussion forums, file sharing, chats, and other activities and materials. Traditional LMSs, in contrast to the present invention, place more of an emphasis on managing individuals involved in problem- or project-based learning processes inside social networks than they do on managing enrollment and content as shown in Figure 7.2. In contrast to enrolling in a course, such technologies are unable to create learning relationships between individuals. Such systems are unable to create an explicit learning relationship indicator. Such systems are unable to keep track of and offer feedback on interactions between members of a social network who have formed a connection of learning.

The majority of social media sites are set up for self-expression and are not used as hubs for organizing group learning activities. Although many of the tools used for social networking, such as wikis, weblogs, and profiles, are easily accessible, there are currently no social networks with embedded curricula or that expressly target online learning with these tools.

7.7 CONCLUSION

Traditionally learning is managed to track individuals' learning and performance. The new system of LMS with tools from social networking can provide an explicit learning environment, not only for online engagement, but to create relationships among the learners for maximum participation to gain from each other's practical knowledge even in remote locations. The models developed here in this process are used to identify and quantify a learner's experience, the e-Learner Experience Model (e-LEM) and the e-Learning Capability Maturity Model (e-LCMM), the latter of which is a model of a technologically advanced e-learning process that help in high engagement and collaboration of student's learning activity and maximize the usage of online services for education purpose.

REFERENCES

I. T. Awidi, "Elements for improving students learning experience in a digital environment." *Workshop presentation on Models for Improving Learning and learning design in e-learning implementation,* University of Twente (November 2005). University of Ghana (May/June 2006).

D. Bateman, & J. Willems, "Facing off: Facebook and Higher Education," in *Misbehavior Online in Higher Education Cutting-edge Technologies in Higher Education,* L. A. Wankel & and C. Wankel, Eds. Emerald, vol. 5, 2012, pp. 53–79. 10.1108/S2044-9968(2012)0000005007

M. Crotty. *The Foundations of Social Research: Meaning and Perspective in the Research Process.* London: SAGE Publications Inc., 1998.

L. Deng, & N. J. Tavares, "From Moodle to Facebook: Exploring students' motivation and experiences in online communities," *Computers and Education,* vol. 68, pp. 167–176, 2013. 10.1016/j.compedu.2013.04.028

M. A. Halliday, "Towards a language-based theory of learning," *Linguistics and Education,* vol. 5, pp. 93–116, 1993. 10.1016/0898-5898(93)90026-7

M. Henderson, N. Selwyn, & R. Aston, "What works and why? Student perceptions of 'useful' digital technology in university teaching and learning," *Studies in Higher Education,* vol. 42, pp. 1–13, 2015. 10.1080/03075079.2015.1007946

B. Jong, C.-H. Lai, Y.-T. Hsia, T.-W. Lin, and Y.-S. Liao, "An exploration of the potential educational value of Facebook," *Computers in Human Behavior.* vol. 32. pp. 201–211, 2014. 10.1016/j.chb.2013.12.007

R. Junco, "Comparing actual and self-reported measures of Facebook use," *Computers in Human Behavior,* vol. 29(3), pp. 626–631, 2013. https://doi.org/10.1016/j.chb.2012.11.007

M. Kent & M. Taylor, "Problems with social media in public relations: Misremembering the past and ignoring the future," *International Journal of Interdisciplinary Research,* vol. 3(2), pp. 23–37, 2014.

S. Manca & M. Ranieri, "Is it a tool suitable for learning? A critical review of the literature on Facebook as a technology-enhanced learning environment," *Journal of Computer Assisted Learning,* vol. 29, pp. 487–504, 2013. 10.1111/jcal.12007

N. Selwyn, "Faceworking: Exploring students' education-related use of Facebook," *Learning. Media and Technology,* vol. 34, pp. 157–174, 2009. 10.1080/1743988090292362

S. Stone, & A. Logan, "Exploring students' use of the social networking site WhatsApp to foster connectedness in the online learning experience," *Irish Journal of Technology Enhanced Learning,* vol. 3, pp. 42–55, 2018. 10.22554/ijtel.v3i1.28

L. S. Vygotsky, *Mind in Society: The Development of Higher Psychological Processes.* Massachusetts, MA: Harvard University Press, 1978.

L. S. Vygotsky, *Thought and Language.* Cambridge, MA: MIT Press, 1986.

Digitization of Industrial Processes

8 Analyzing Consumer Product Feedback Dynamics with Confidence Intervals

*A. Venkata Subramanian, Vithya Ganesan,
and Viswanathan Ramasamy*

8.1 INTRODUCTION

Sentimental classification and sentimental analysis are the subsets of text analysis and classification. To identify positive, negative, or neutral sentiment a phrase or a group of words is analyzed for classifying/grouping sentiments. It is helpful in sentiment analysis of product feedback, questionnaire, and answer analysis. Analyzing sentiment across product feedback, questionnaires, and answers provides a comprehensive understanding. Comparing sentiment analysis from company questions and answers with product reviews enhances acknowledgment and authentication of consumer comments regarding the product.

For example, Amazon's web pages collect customer feedback about products by floating question answers and reviews. Take for example a customer who may purchase an iPhone 13, iPhone 13 Pro, or iPhone 13 Pro Max, etc. Compare customer reviews and questions and answers for the iPhone 13, iPhone 13 Pro, and iPhone 13 Pro Max using the BERT neural network

8.2 LITERATURE SURVEY

E-commerce has recently experienced rapid growth. As a result, online purchasing has increased, which has resulted in an increase in online product reviews. Because the customer's opinion about the product is influenced by other consumers' recommendations or complaints, the opinions in customer reviews have a massive influence on customer's purchasing decisions. This study analyzes the Amazon reviews data set and investigates sentiment classification using questions and answers and reviews.

E-commerce remark sentiment analysis has become a research focus. The current word vector representation model does not take into account the text context [1].

DOI: 10.1201/9781003388241-13

As a result, the SA-BERT pre-training language model based on transformer bidirectional encoder representation is proposed in this study [2–5]. BERT first encodes the word vector to reflect the semantic information of the remark text's context. The attention method is then utilized to extract text characteristics at a deeper level, grasp the semantics of text information, and finish the e-commerce comment sentiment analysis assignment [6].

In this paper, we have scraped the reviews, questions and answers of the product from the Amazon website because from the customers opinions only the sentiment of the product can be analyzed. This web scraping can be achieved using the BeautifulSoup package where it pulls out the required data from all HTML or XML files.

We read reviews, and questions and answers about the product and searched for the data we needed on the web and then did our emotional analysis [7].

For comparing consumer/customer doubt and product legitimacy, the BERT Neural Network and Python are used to look at the sentimental results of Amazon questions and answers, Amazon reviews of a product, and make a graph of them. The following types of data were used to compare what customers had to say [8–15].

8.3   DATA SET

In this data set, we have labels like questions, answers, sentiment_question, sentiment_answers, reviews-title, sentiment_titles, reviews, sentiment_reviews, as shown in Figure 8.1.

Labels:

- Questions: The pre-processed, scraped questions data on the particular product.
- Answers: The pre-processed, scraped answers data on the particular product against the questions.
- Sentiment_question: After performing sentimental analysis on the question data, it gave us ratings from 1-5, which indicate poor, unsatisfactory, satisfactory, good, and outstanding, respectively.
- Sentiment_answers: After performing sentimental analysis on the question data, it gave us ratings from 1-5, which indicate poor, unsatisfactory, satisfactory, good, and outstanding, respectively.
- Reviews_title: The title of reviews on the product which is given by the customer.
- Sentiment_titles: The rating on reviews titles where we perform sentimental analysis on reviews titles.
- Reviews: Detailed view of the full review of the customer on the product.
- Sentiment_reviews: The sentimental analysis result on the reviews data.

8.4   DATA PRE-PROCESSING

The data set utilized had to be pre-processed before use to make sure that the operation was completed as effectively as feasible. First we pre-processed the data of questions on the product, as shown in Figure 8.2.

	questions	answers	sentiment_question	sentiment_answers	reviews_title	sentiment_titles	reviews	sentiment_reviews
0	what all accessories comes in the box with phone?	just the usb c to lightning cable and the sim removing tool	1	5	Do not purchase this I phone	1	this product is duplicate of iphone 12 only camera design changed. very low quality of this product buy samsung mobile phone	1
1	is this model.how much capacity of ram?	6gb	3	5	😊😊😊😊	5	you should be a bigggggg bot to buy this phoneyou could get iphone 12 pro that's the best	5
2	what is the camera?	12 mp!	1	1	Damaged product received	1	there was a hairline scratch on the screen which is not clearly visible directly. when we tried contacting amazon for the replacement, they asked us to connect to apple and when connected to apple, they asked to connect with amazon. blame game begins and we end up paying the price. awfull service. i am going to tag on the social media both amazon and apple. such a big scam	1
3	it camera is very clear	yes best camera available	4	5	Excellent upgrade in context of being 90k poorer.	5	just but iphone 12!it's not worth the money!unless ur from bjp or something where you're full with corrupt money, then go ahead.	1
4	does it still have a notch?	yes	3	5	trash af	1	i am poor and i cant afford it lmao xdxdxdxdxd so if you have money to flex buy it i am bored thats why i am writing this review while drinking coffee i have my chemistry exam on manday and cant study mole concept and got bored thats why i came here to write this review if you read this review til here very good you justt wasted your time now go and do something productive :)	1

FIGURE 8.1 Final data set.

FIGURE 8.2 Questions before data pre-processing.

```
Out[284]:  ['what all accessories comes in the box with phone?',
            'is this model,how much capacity of ram?',
            'what is the camera?',
            'it camera is very clear',
            'does it still have a notch?',
            "hi friends i want to buy a mobile. but confused between iphone13 promax and samsung galaxy s22ultra. present i'm
            using samsung galaxy note 8.plz help.",
            'this phone 5g',
            'iphone 13 pro max sierra blue 128',
            'how many battery',
            'this phone is dual sim or single sim',
            'unable to place order, when will this get back in stock?',
            'if i buy this which thing will come first sadness or happiness ?',
            'exchange is available',
            '13 pro max is what model is this',
            'does this model support 90 fps in bgmi?',
            'is it unlocked for all the carrier in the world...??',
            'my order of iphone will come tomorrow can i change to 256',
            'what is in packing',
```

FIGURE 8.3   Questions after data pre-processing.

```
Out[7]:  ['Just the USB C to Lightning Cable and the SIM Removing Tool',
          '6GB',
          '12 MP!',
          'Yes best camera available',
          'Yes',
          'I have an I phone but I prefer android over ios . Simplicity is more important than security for me',
          'Yes',
          'yes',
          '1',
          'dual one physical and one e sim',
          'Yes',
          'happiness followed by sadness',
          "Exchange of old mobile is available in most of the pin codes... Even if you get the product as damaged (mostly it
          won't happen) they will replace the product, only after a clear check.",
          'Very nice',
          'Yes it supports',
          'Yes',
          'No, you have to cancel it / reject the deliver and pleace a new request/order for the 256 one!',
          'Mobile and Charging cable.',
```

FIGURE 8.4   Answers after data pre-processing.

Then we pre-processed the data of answers for the product, which are given to the questions shown in Figures 8.3 and 8.4.

8.5   PROPOSED SYSTEM

To web-scrape data from a website like Amazon, specifically targeting reviews, questions, and answers of a particular product, you can use Python along with libraries like BeautifulSoup and requests. From this segmentation step, we pre-processed the data which is in three segments.

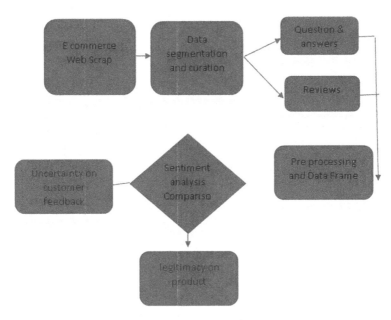

FIGURE 8.5 Trajectory of customer feedback on product.

This step included cleaning the unwanted data, e.g., removing emojis, filling null values, etc. After the data pre-processing step, the cleaned data was converted into three data frames, which refer to three segments. Figure 8.5 shows the consumer feedback and its sentiment analysis.

8.6 IMPLEMENTATION

Transformers offer hundreds of pre-trained models for performing tasks on many modalities such as text, vision, and audio. These models may be used on text in over 100 languages for tasks such as text categorization, information extraction, question answering, summarization, translation, and text production.

From the Transformers module, we imported some libraries like Auto Tokenizer and AutoModelForSequenceClassification. This Auto Tokenizer library helps in creating a class directly. The AutoModelForSequenceClassification library helps in training the model.

Second, we imported torch, requests, and RE libraries in Python, where the torch is a Tensor library similar to NumPy, with robust GPU support. PyTorch has a unique way of building neural networks: using and replaying a tape recorder. The requests library helps to allow one to send HTTP/1.1 requests extremely easily. Certainly, nowadays, it's more common and recommended to use the json method for forming encoded POST data. This approach simplifies the process and is more intuitive. The RE library assists us in defining a collection of strings that match it; the methods in this module allow you to determine whether a certain string matches a given regular expression (or if a given regular expression matches a particular string, which comes

	questions	answers	sentiment_question	sentiment_answers	reviews_title	sentiment_titles	reviews	sentiment_reviews
0	what all accessories comes in the box with phone?	just the usb c to lightning cable and the sim removing tool	1	5	Do not purchase this l phone	1	this product is duplicate of iphone 13 only camera design changed. very low quality of this product buy samsung mobile phone	1
1	is this model,how much capacity of ram?	6gb	3	5	🙂🙂🙂🙂	5	you should be a bigggggg bot to buy this phoneyou could get iphone 12 pro that's the best	5
2	what is the camera?	12 mp!	1	1	Damaged product received	1	there was a hairline scratch on the screen which is not clearly visible directly. when we tried contacting amazon for the replacement, they asked us to connect to apple and when connected to apple, they asked to connect with amazon. blame game begins and we end up paying the price. awfull service. i am going to tag on the social media both amazon and apple. such a big scam	1
3	it camera is very clear	yes best camera available	4	5	Excellent upgrade in context of being 90k poorer.	5	just but iphone 12!it's not worth the money!unless ur from bip or something where you're full with corrupt money, then go ahead.	1
4	does it still have a notch?	yes	3	5	trash af	1	i am poor and i cant afford it lmao xdxdxdxdxd so if you have money to flex buy it i am bored thats why i am writing this review while drinking coffee i have my chemistry exam on manday and cant study mole concept and got bored thats why i came here to write this review if you read this review til here very good you justt wasted your time now go and do something productive :)	1

FIGURE 8.6   Final data set after sentiment analysis.

down to the same thing). Regular expressions can be concatenated to create new ones; for example, if A and B are both regular expressions, then AB is also a regular expression. If the string p matches A and the other string q matches B, the string pq will match AB. This is true unless A or B have low priority operations, boundary conditions between A and B, or numbered group references. As a result, complicated expressions may be easily built from simpler primitive expressions. Figure 8.6 shows the data set after sentiment analysis.

Third, from the bs4 module, BeautifulSoup package was imported. It is used for web-scraping purposes to pull the data out of HTML and XML files. It creates a parse tree from page source code that can be used to extract data in a hierarchical and more readable manner.

Questions about goods on Amazon were gathered from its URLs and were put in a list. We cleaned both the answers and the list of questions, and then we converted them into two data frames. Then we undertook to concat those two data frames into a single one as the corresponding questions get their corresponding answers, and get a sentimental rating for both questions and answers separately from the 1-5 rating (where 1=poor, 2=unsatisfactory, 3=satisfactory, 4=good, 5=outstanding). According to those analyses, we will count each rating for questions and answers and we will visualize them.

8.7   RESULTS

The confidence interval was calculated for questions and answers and reviews, as shown in the Figures 8.7 and 8.8.

	mean	count	std		
sentiment_question					
1	3.111111	45	1.760969		
2	3.000000	12	1.758098		
3	3.804878	41	1.520189		
4	4.375000	8	0.916125		
5	3.457143	35	1.668794		
· ·					
	mean	count	std	ci95_hi	ci95_lo
sentiment_question					
1	3.111111	45	1.760969	3.625630	2.596592
2	3.000000	12	1.758098	3.994738	2.005262
3	3.804878	41	1.520189	4.270209	3.339547
4	4.375000	8	0.916125	5.009843	3.740157
5	3.457143	35	1.668794	4.010015	2.904271

FIGURE 8.7   Confidence interval of sentiment question.

	mean	count	std		
sentiment_reviews					
1	3.782609	23	1.704447		
2	3.750000	12	1.602555		
3	4.000000	11	1.549193		
4	4.250000	40	1.276011		
5	3.890909	55	1.486641		
· ·					
	mean	count	std	ci95_hi	ci95_lo
sentiment_reviews					
1	3.782609	23	1.704447	4.479196	3.086021
2	3.750000	12	1.602555	4.656731	2.843269
3	4.000000	11	1.549193	4.915515	3.084485
4	4.250000	40	1.276011	4.645440	3.854560
5	3.890909	55	1.486641	4.283808	3.498010

FIGURE 8.8   Confidence interval of sentiment reviews.

Figures 8.9, 8.10, 8.11, 8.12, and 8.13 show the visualization for questions and reviewer sentiment analysis.

Figure 8.14 shows the confidence interval between the question answer and reviews of consumer/customer feedback on products. When % is high, it shows the legitimacy of consumer feedback. If it is less, it shows ambiguity in the consumer's mind and the product. Figure 8.15 shows subplot representation for sentiment analysis.

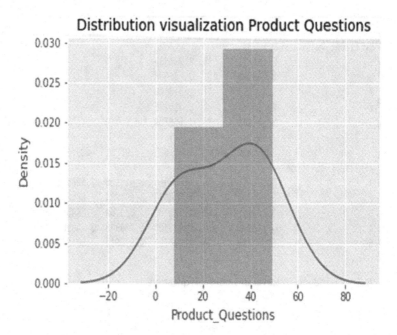

FIGURE 8.9   Distribution visualization for product questions.

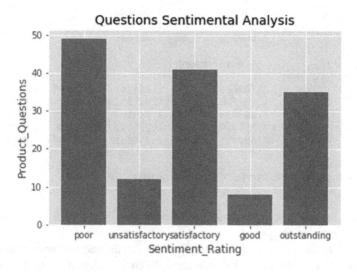

FIGURE 8.10   Questions sentimental analysis.

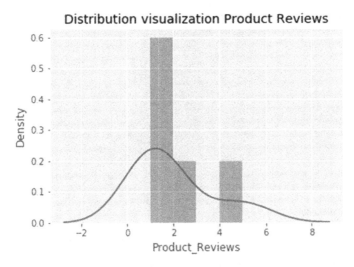

FIGURE 8.11   Distribution visualization product reviews.

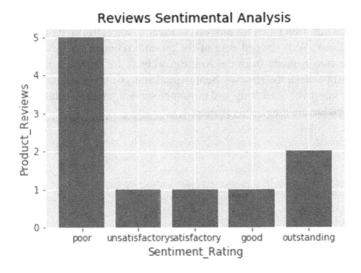

FIGURE 8.12   Reviews sentimental analysis.

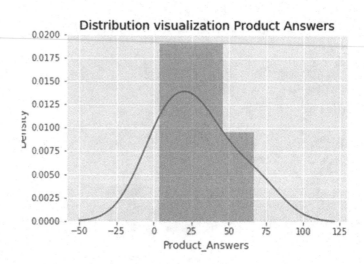

FIGURE 8.13 Distribution visualization product answers.

8.8 CONCLUSION

To conclude, the goal of the proposed research is to solve the semantic analysis of consumer/customer feedback which is based upon their opinions on the product. Web-scraped data has been taken for analysis to identify the authenticity of consumer/customer feedback. Web-scraped data of the product iPhone13 reviews, questions and answers section is taken from the Amazon website for comparison and to identify the trade-off between the reviews. Sentimental analysis was undertaken using BERT. This study improves marketing and customer service, and this strategy is helpful to the policy maker and marketing planner.

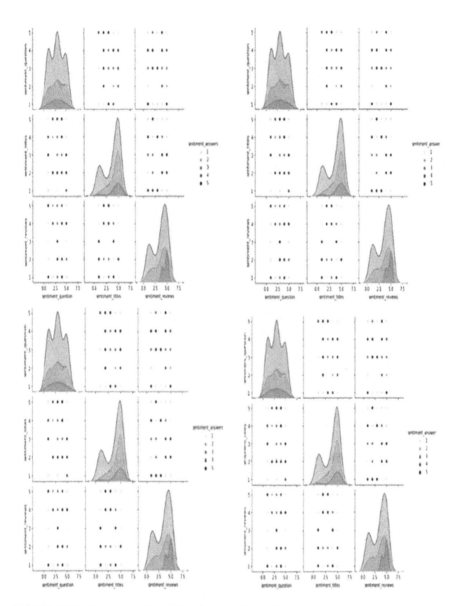

FIGURE 8.14 Representation of data frame with subplot.

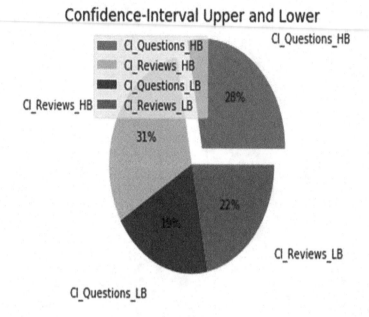

FIGURE 8.15   Confidence interval of product question and reviews.

REFERENCES

1. Arwa, S. M., and AlQahtani, (2021). "Product Sentiment Analysis for Amazon Reviews." 13(3). doi: 10.5121/ijcsit.2021.13302 http://aircconline.com/ijcsit/V13N3/13321ijcsit02.pdf
2. Bai, X. (2011). "Predicting Consumer Sentiments from Online Text." Decis Support Syst 50(4):732–742.
3. Cao, Q., Duan, W., and Gan, Q. (2011). "Exploring Determinants of Voting for the "Helpfulness" of Online User Reviews: A Text Mining Approach." Decis Support Syst 50(2):511–521.
4. Glez-Peña, D., Lourenço, A., López-Fernández, H., Reboiro-Jato, M., and Fdez-Riverola, F. (2013, May). "Web Scraping Technologies in an API world," Briefings in Bioinformatics 15(5):788–797. doi: 10.1093/BIB/BBT026
5. Gothai, E., Muthukumaran, V., Valarmathi, K., Sathishkumar, V. E., Thillaiarasu, N., & Karthikeyan, P. (2022). "Map-Reduce Based Distance Weighted k-Nearest Neighbor Machine Learning Algorithm for Big Data Applications." Scalable Computing: Practice and Experience 23(4):129–145.
6. Kaladevi, A. C., Saravanakumar, R., Veena, K., Muthukumaran, V., Thillaiarasu, N., & Kumar, S. S. (2022). "Data Analytics on Eco-Conditional Factors Affecting Speech Recognition Rate of Modern Interaction Systems." Journal of Mobile Multimedia, 2, 1153–1176.
7. Liu, Y., Lu, J., Yang, J., and Mao, F. (2020). "Sentiment Analysis for E-Commerce Product Reviews by Deep Learning Model of Bert-BiGRU-Softmax." 17, no. July:7819–7837. doi: 10.3934/mbe.2020398

8. "nlptown/bert-Base-Multilingual-Uncased-Sentiment Model - NLP Hub - Metatext." https://huggingface.co/nlptown/bert-base-multilingual-uncased-sentiment

9. Preethi, P., Asokan, R., Thillaiarasu, N., & Saravanan, T. (2021). "An Effective Digit Recognition Model Using Enhanced Convolutional Neural Network Based Chaotic Grey Wolf Optimization." Journal of Intelligent & Fuzzy Systems 41(2):3727–3737.

10. "Scrape Beautifully with Beautiful Soup in Python - Tutorial." https://analyticsindiamag.com/beautiful-soup-webscraping-python/

11. Shenoy, A., and Thillaiarasu, N. (2022). "A Survey on Different Computer Vision Based Human Activity Recognition for Surveillance Applications." 2022 6th International Conference on Computing Methodologies and Communication (ICCMC). IEEE.

12. Shyamambika, N., and Thillaiarasu, N. (2016). "A Survey on Acquiring Integrity of Shared Data with Effective User Termination in the Cloud." 2016 10th International Conference on Intelligent Systems and Control (ISCO). IEEE.

13. Srilatha, D., and Thillaiarasu, N. (2022). "DDoSNet: A Deep Learning Model for detecting Network Attacks in Cloud Computing." 2022 4th International Conference on Inventive Research in Computing Applications (ICIRCA). IEEE.

14. Srilatha, D., and Thillaiarasu, N. (2022, September). "OIDCBMS: A Novel Neural Network based Intrusion Detection System to Cloud Computing based on New Cube Algorithm." In 2022 4th International Conference on Inventive Research in Computing Applications (ICIRCA) (pp. 1651–1656). IEEE.

15. Srilatha, D., and Thillaiarasu, N. (2023). "Implementation of Intrusion Detection and Prevention with Deep Learning in Cloud Computing." Journal of Information Technology Management 15. Special Issue: 1–18.

16. Thillaiarasu, N., Pandian, S. C., Vijayakumar, V., Prabaharan, S., Ravi, L., & Subramaniyaswamy, V. (2021). Designing a Trivial Information Relaying Scheme for Assuring Safety in Mobile Cloud Computing Environment. Wireless Networks 27:5477–5490.

17. Thillaiarasu, N., and ChenthurPandian, S. (2016). "Enforcing Security and Privacy over Multi-Cloud Framework Using Assessment Techniques." 2016 10th International Conference on Intelligent Systems and Control (ISCO). IEEE.

18. Xie, S., Cao, J., Wu, Z., Liu, K., Tao X., and Xie, H. (2020). "Sentiment analysis of chinese e-commerce reviews based on BERT," 2020 IEEE 18th International Conference on Industrial Informatics (INDIN) (pp. 713–718). doi: 10.1109/INDIN45582.2020.9442190

Information System in Industry

9 Amplifying the Effectiveness of a Learning Management System

Exploring the Impact of NEP-Compliant Curriculum Changes on Higher Education Institutions

Dara Vijaya Lakshmi and Chitra Kesavan

OBJECTIVES

- To assess the current state of the learning management system (LMS) effectiveness in higher education institutions (HEIs) and identify areas for improvement.
- To examine the principles and guidelines outlined in the National Education Policy (NEP) and their relevance to curriculum development and LMS integration.
- To explore the perceptions and experiences of administrators, faculty members, and students regarding the integration of NEP-compliant curriculum changes into LMS platforms.
- To investigate the impact of NEP-compliant curriculum changes on LMS usage, student engagement, and academic performance within HEIs.
- To identify the benefits, challenges, and potential areas for improvement in implementing NEP-driven curriculum modifications in LMS platforms.
- To determine the correlation between NEP compliance, LMS effectiveness, and institutional outcomes, such as student performance and retention rates.
- To provide evidence-based insights and recommendations for policymakers, administrators, and educators on leveraging NEP-compliant curriculum changes to enhance LMS effectiveness in HEIs.
- To contribute to the existing literature on educational technology and curriculum development by examining the effects of policy-driven modifications on LMS platforms.

DOI: 10.1201/9781003388241-15

By addressing these objectives, this study aims to provide a comprehensive understanding of the impact of NEP compliant curriculum changes on LMS effectiveness within HEIs.

9.1 INTRODUCTION

The effectiveness of learning management systems (LMS) in higher education institutions (HEIs) has become increasingly crucial in the ever-evolving educational landscape. As advancements in technology continue to shape the way students learn and educators teach, it is imperative to explore ways to enhance LMS platforms to meet the changing needs of the educational environment. One such avenue for improvement lies in aligning the curriculum with NEP (National Education Policy) guidelines, which advocate for comprehensive educational reforms.

The National Education Policy serves as a framework for educational development, emphasizing holistic and multidisciplinary learning experiences, critical thinking skills, and student engagement. By integrating NEP-compliant curriculum changes into LMS platforms, HEIs can potentially leverage the benefits of these policy-driven modifications to enhance the effectiveness of their educational systems.

This study aims to investigate the impact of NEP-compliant curriculum changes on LMS effectiveness within HEIs. By examining the effects of these changes on LMS usage, student engagement, and academic performance, this research seeks to provide valuable insights into the potential benefits and challenges associated with aligning the curriculum with NEP guidelines.

To achieve this objective, a mixed-methods approach combining qualitative and quantitative techniques will be employed. The qualitative phase of the study will involve in-depth interviews with administrators, faculty members, and students to gather their perspectives on the integration of NEP-driven changes into the LMS. These interviews will provide valuable insights into the perceived advantages, challenges, and potential areas for improvement in implementing NEP-compliant curriculum modifications.

Simultaneously, the quantitative phase will involve distributing surveys among a representative sample of HEIs to collect data on LMS usage, student engagement, and academic performance. Statistical analyses, such as regression models and correlation analyses, will be conducted to examine the relationship between NEP compliance, LMS effectiveness, and institutional outcomes.

By exploring the impact of NEP-compliant curriculum changes on LMS effectiveness, this research aims to contribute to the existing literature on educational technology and curriculum development. The findings of this study will provide evidence-based insights that can guide policymakers, administrators, and educators in making informed decisions regarding the integration of NEP-driven changes into LMS platforms.

9.2 SIGNIFICANCE OF THE STUDY

The significance of this study lies in its exploration of the impact of National Education Policy (NEP)-compliant curriculum changes on the effectiveness of Learning

Management Systems (LMS) in higher education institutions. Understanding this impact is crucial for several reasons:

Educational Reform: The NEP represents a significant educational reform aimed at transforming the higher education landscape. By investigating the influence of NEP-aligned curriculum changes on LMS platforms, this study contributes to the ongoing discourse on educational reform and its practical implementation.

Enhancing Learning Outcomes: LMS platforms offer a wide range of tools and features that can enhance student engagement, collaboration, and personalized learning. By examining the impact of NEP-compliant curriculum changes on LMS adoption, this study can provide insights into how institutions can leverage these technologies to improve learning outcomes and foster student success.

Administrative Efficiency: The NEP emphasizes streamlining administrative processes and reducing bureaucratic hurdles. By exploring how NEP-compliant curriculum changes affect the administrative aspects of higher education institutions, this study can identify opportunities to leverage LMS systems for efficient data management, course administration, and assessment processes.

Faculty Adoption and Support: Faculty play a critical role in the successful implementation of curriculum changes and the effective use of LMS platforms. Understanding the challenges and perspectives of faculty members regarding NEP-aligned curriculum changes can help institutions develop targeted strategies to support faculty adoption, provide necessary training, and address potential barriers to implementation.

Institutional Decision-Making: Higher education institutions need evidence-based insights to guide decision-making regarding curriculum changes and investments in educational technologies. By examining the impact of NEP-compliant curriculum changes on LMS effectiveness, this study can provide valuable information to educational leaders, administrators, and policymakers, enabling them to make informed decisions and allocate resources effectively.

Future Research and Development: The findings of this study can stimulate further research and development in the field of educational technology. By identifying the opportunities and challenges associated with NEP-compliant curriculum changes and LMS integration, this study can pave the way for future investigations on innovative pedagogical practices, emerging technologies, and best practices in LMS implementation.

The implementation of NEP-compliant curriculum changes provides an opportunity to address accessibility and inclusivity in higher education. By investigating the impact of these changes on LMS effectiveness, this study can identify how institutions can leverage LMS platforms to accommodate diverse learners, including those with disabilities, different learning styles, and varying educational backgrounds. Understanding the impact of NEP-aligned curriculum changes on LMS adoption can help institutions ensure that educational resources and materials are accessible, adaptable, and inclusive, thus promoting equitable educational opportunities for all students.

Promoting Lifelong Learning and Skills Development: The NEP emphasizes the importance of lifelong learning and the development of critical skills to meet the demands of a rapidly changing world. By exploring the impact of NEP-compliant curriculum changes on LMS effectiveness, this study can shed light on how LMS platforms can be utilized to foster continuous learning, skill development, and professional growth beyond traditional classroom settings. Understanding the potential of LMS systems to offer flexible learning pathways, competency-based assessments, and opportunities for self-paced learning can inform institutions on how to effectively align curriculum changes with the goal of lifelong learning.

Meeting the Needs of Digital Natives: Today's higher education students are digital natives who have grown up in a technologically advanced world. Investigating the impact of NEP-compliant curriculum changes on LMS platforms can provide insights into how institutions can meet the expectations and preferences of these digital natives. Understanding how LMS platforms can support interactive and engaging learning experiences, collaborative work environments, and seamless access to educational resources can help institutions adapt their teaching practices and leverage technology effectively to meet the needs and preferences of modern learners.

Addressing Technological Challenges: Implementing NEP-compliant curriculum changes and integrating LMS platforms require institutions to navigate various technological challenges. This study can help identify common technological barriers and provide recommendations for overcoming them. By addressing issues such as infrastructure requirements, compatibility with existing systems, data security, and technical support, institutions can better plan and allocate resources to ensure a smooth transition and maximize the benefits of NEP-compliant curriculum changes in conjunction with LMS integration.

9.3 REVIEW OF LITERATURE

Learning Management Systems in Higher Education:

Learning Management Systems (LMS) have become integral to higher education institutions as they offer a centralized platform for managing and delivering educational content, facilitating communication, and assessing student progress. LMS platforms provide a range of tools and features such as course management, content creation and sharing, discussion forums, assignment submission, and grading. Numerous studies have explored the benefits of LMS adoption, including increased student engagement, improved learning outcomes, and enhanced administrative efficiency (Graham, 2021).

9.3.1 NATIONAL EDUCATION POLICY AND CURRICULUM CHANGES

The National Education Policy (NEP) represents a comprehensive educational reform initiative aimed at transforming the Indian higher education system. The NEP emphasizes a learner-centric approach, multidisciplinary education, flexibility in curriculum design, and the integration of technology in teaching and learning. These policy changes aim to foster critical thinking, creativity, and employability skills among students (Aktar, 2021).

The Role of LMS in Implementing NEP-Compliant Curriculum:

NEP-compliant curriculum changes require institutions to rethink their teaching methodologies and leverage technology effectively. LMS platforms can play a crucial role in implementing NEP-aligned curriculum by offering tools for interactive and collaborative learning, personalized learning pathways, competency-based assessments, and flexible delivery modes (Foreman et al., 2021; Yousafzai et al., 2022). LMS platforms can facilitate the adoption of innovative pedagogical approaches, such as flipped classrooms, blended learning, and active learning strategies (Yang et al., 2016).

9.3.2 Challenges and Opportunities of NEP-Aligned Curriculum on LMS Platforms

The integration of NEP-compliant curriculum changes with LMS platforms presents both challenges and opportunities for higher education institutions. Challenges include faculty resistance to change, lack of technical skills and training, concerns about privacy and data security, and the need for adequate infrastructure and support systems (Ho et al., 2017; Ng et al., 2020). However, implementing NEP-compliant curriculum changes on LMS platforms also offers opportunities for enhancing student engagement, promoting active learning, fostering collaboration, and enabling lifelong learning (Hussain et al., 2018).

9.3.3 Current Gaps and Research Directions

While some studies have explored the individual impacts of LMS adoption and NEP-compliant curriculum changes, there is a need for more research on the combined effect of these factors in higher education institutions. Future studies can investigate the relationship between NEP-aligned curriculum changes and LMS effectiveness by examining student learning outcomes, faculty perspectives and experiences, institutional readiness and support, and the influence on administrative processes (Yousafzai et al., 2022). Additionally, research can focus on identifying effective strategies for faculty development, addressing technological challenges, and ensuring equitable access and inclusivity in LMS-based learning environments.

The review would likely include studies that have examined the impact of curriculum changes aligned with NEP guidelines on LMS usage, student engagement, and academic performance. It would explore various dimensions of LMS effectiveness, such as the adoption and utilization of LMS platforms, student satisfaction and motivation, learning outcomes, and institutional performance indicators.

The empirical review might reveal studies that have employed both quantitative and qualitative research methods. Quantitative studies could involve survey-based approaches, experimental designs, or statistical analyses to measure the relationship between NEP-compliant curriculum changes and LMS effectiveness. Qualitative studies, on the other hand, could provide insights into the perceptions, experiences, and challenges encountered during the integration of NEP-driven changes into LMS platforms.

9.4 RESEARCH METHODOLOGY

9.4.1 Research Design

The research design for this study will involve a mixed-methods approach, combining quantitative and qualitative data collection and analysis techniques. This approach allows for a comprehensive exploration of the impact of NEP-compliant curriculum changes on the effectiveness of Learning Management Systems (LMS) in higher education institutions.

9.4.2 Data Collection

a. Quantitative Data: Quantitative data will be collected through surveys administered to students, faculty members, and administrators from multiple higher education institutions. The survey will include Likert scale questions and closed-ended items to assess factors such as LMS adoption, student engagement, learning outcomes, administrative efficiency, and faculty perspectives. Demographic information and institutional characteristics will also be collected.
b. Qualitative Data: Qualitative data will be collected through in-depth interviews and focus group discussions with faculty members, administrators, and LMS administrators. These interviews will provide a deeper understanding of the challenges, opportunities, and experiences related to the implementation of NEP-compliant curriculum changes on LMS platforms. The qualitative data will be audio-recorded and transcribed for analysis.

9.4.3 Data Analysis

a. Quantitative Data Analysis: The quantitative data collected from the surveys will be analyzed using descriptive statistics to examine the frequencies, means, and standard deviations of the responses. Inferential statistical techniques, such as correlation analysis and regression analysis, will be employed to explore relationships between variables and determine the impact of NEP-compliant curriculum changes on LMS effectiveness.
b. Qualitative Data Analysis: The qualitative data obtained from interviews and focus group discussions will be analyzed using thematic analysis. The transcripts will be coded, and themes and patterns related to the impact of NEP-compliant curriculum changes on LMS platforms will be identified. These themes will be interpreted and presented to provide a comprehensive understanding of the experiences, challenges, and opportunities associated with the integration of NEP-compliant curriculum changes and LMS.

9.4.4 Presentation and Interpretation of Data

The first question on the questionnaire aimed to assess the level of innovation within educational management. The findings, summarized in Table 9.1, provide insights into participants' perceptions of innovation creation in educational management.

The data analysis revealed that 41% of respondents acknowledged a high level of innovation creation within educational management. Furthermore, 19.68% of participants considered the level of innovation creation to be extremely high. These percentages indicate a positive trend toward innovation in educational management, suggesting that a considerable portion of the respondents perceive a significant degree of quality and advancement at the secondary school level.

However, the data also revealed that in some institutions, innovation creation is infrequent (26.22%) or nonexistent (13.11%). This suggests that there are certain institutions where decision-makers and/or academic personnel may lack confidence, display conservative tendencies, or exhibit a lack of motivation towards innovation.

The presence of institutions with limited innovation creation could be attributed to several factors. Decision-makers and academic personnel in these institutions may face challenges in embracing change, be resistant to new ideas, or encounter barriers to implementing innovative practices. Additionally, limited resources, lack of training opportunities, or a rigid organizational culture may hinder the creation of innovation in these institutions.

To foster a culture of innovation within educational management, it is important to address the barriers identified in institutions where innovation creation is infrequent or absent. Providing professional development opportunities focused on innovation, promoting a supportive and open organizational culture, and encouraging collaboration and knowledge-sharing among decision-makers and academic personnel can help overcome these challenges. Moreover, showcasing successful instances of innovation within educational management and highlighting the positive outcomes it brings can serve as a motivator for change and inspire individuals and institutions to embrace innovative practices.

Overall, the findings suggest a positive trend towards innovation creation in educational management, although there are still institutions where improvement is needed. By identifying the challenges and opportunities associated with innovation creation, educational leaders and policymakers can develop strategies to foster a culture of innovation and drive continuous improvement in educational management practices.

The second item on the questionnaire aimed to assess the level of acceptance or resistance to change within the school organization. The data analysis, presented in Table 9.2, provides insights into participants' attitudes towards change and its relationship to institutional effectiveness.

The findings indicate a high level of acceptance of change within the school organization, with 42.62% of respondents acknowledging a high level of acceptance, and 39.34% perceiving a very high level of acceptance. These percentages demonstrate that a significant portion of the participants recognize the importance of change in driving innovation and improving institutional effectiveness.

By comparing the results from Tables 9.1 and 9.2, it is evident that there is a synchronization of findings, indicating a clear dependency between change and the

TABLE 9.1
The Degree of Innovation Production

Item	Very high degree	High degree	Low degree	None at all	Total
State the degree in which innovation is produced in your institution	12 19.68%	25 41.00%	16 26.22%	8 13.11%	61 100%

TABLE 9.2
Accepting Change

Item	A very high level	A high level	A low level	Not at all	Total
At which level is change accepted in your organization?	24 39.34%	26 42.62%	9 14.75%	2 3.28%	61 100%

generation of innovative ideas. The positive correlation between the acceptance of change and innovation creation suggests that embracing change is necessary to foster a culture of innovation within the school organization.

On the other hand, a small proportion of organizations (14.75%) displayed low acceptance of change, indicating resistance to making even minor changes. The reasons for this resistance may be attributed to various factors. Managerial reasons, such as wrong decisions, rigid procedures, and difficulties in implementation, may hinder the acceptance of change. Insufficient information or incomplete explanations about the need for change, economic-financial constraints, subjective reasons like lack of motivation or the expectation of losing privileges, and managerial resource limitations could also contribute to resistance.

While there are instances where change is rapid, these occurrences seem to be decreasing in frequency. This decline in the number of rapid changes suggests the need for a more proactive approach to change management within the school organization.

To address resistance to change and promote a culture of acceptance, it is crucial for educational leaders and administrators to provide clear communication about the need for change, its benefits, and the expected outcomes. Engaging stakeholders in the change process, providing adequate resources and support, and fostering a climate of trust and collaboration can help overcome resistance and encourage a more positive attitude towards change, the findings indicate a high level of acceptance of change within the school organization, aligning with the importance of change for innovation generation. However, a proportion of organizations still exhibit resistance to change. By understanding the reasons for resistance and implementing effective change management strategies, educational leaders can foster a culture of acceptance and drive institutional effectiveness through continuous improvement and innovation.

In another section of the questionnaire, the characteristics of pedagogical techniques that contribute to the growth of educational institutions were explored. The findings, presented in Table 9.3, shed light on participants' perspectives on effective strategies for institutional growth.

According to the centralized responses, one method valued by 26.22% of respondents for institutional growth is consensus and responsible action. This suggests that stakeholders recognize the importance of collaborative decision-making and taking responsibility for actions as key drivers of institutional growth.

The second factor highlighted by 24.60% of respondents is the functioning of educational relationships between colleges, families, and the community. This perspective

TABLE 9.3
Effective Means of Developing the Educational Institution

Item	N	%
Application of innovative strategies	11	18.03%
Consensus and responsible action	16	26.22%
Careful use of material and financial strategies	6	9.83%
Functioning of school-parent-community partnership	15	24.60%
Total	61	100%

reflects the pragmatism in understanding that institutional growth is successful when there is active collaboration and engagement with external stakeholders.

A notable portion of respondents (21.31%) agrees that motivating the staff and school community is crucial for institutional development. This finding emphasizes the interdependence between rising levels of motivation among students and teaching staff and the desired growth of the institution. It underscores the importance of creating an environment that fosters motivation and engagement to drive institutional progress.

Although ranked last on respondents' lists (18.03%), the use of innovative tactics was still recognized as a successful approach to growing educational institutions. This indicates an understanding of the role of innovation in educational management, aligning with contemporary socio-educational realities. The recognition of the value of innovative tactics highlights the need to embrace emerging educational trends and practices for sustainable institutional growth.

In terms of resource management, careful use of both material and financial resources ranked last with 9.83% of the vote. This position can be interpreted as a reflection of the evolving role of educational managers. The focus has shifted from being excellent managers of resources to becoming coordinators and initiators of transparent methods that optimize resource allocation, the findings underscore the importance of consensus and responsible action, educational relationships, staff and community motivation, innovative tactics, and resource management for institutional growth. These perspectives provide valuable insights into effective strategies for fostering the development of educational institutions. By understanding these factors and incorporating them into decision-making and planning processes, educational leaders can cultivate a conducive environment for growth and advancement, ensuring the long-term success of the institution.

College managers are faced with the need to take action to compete in the educational market, driven by objective factors and constraints both within and outside of colleges. The primary academic action directions identified by respondents in Table 9.4 are aimed at achieving this goal.

According to 31.15% of managers, increased institutional autonomy is viewed as the key to achieving a high degree of competitiveness on the educational market. This option aligns with educational policies that support decentralization and assume local-level responsibility for institutional decision-making. The emphasis on institutional

TABLE 9.4

Managerial Action Directions in View of Becoming Competitive on the Educational Market

Item	N	%
Implementation of practices, structures, new and innovative techniques	13	21.31%
Production of a quality culture	6	9.83%
Increase in institutional autonomy	19	31.15%
Development and application of educational strategies	13	21.31%
Increase in quality of initial and continuing training	10	16.39%
Total	61	100%

autonomy can be justified as it enables colleges to adapt quickly to market demands and make timely decisions that enhance their competitiveness.

The second-ranked direction, highlighted by 21.31% of respondents, focuses on the creation and utilization of educational marketing strategies, as well as the adoption of cutting-edge procedures, frameworks, and methods. This finding indicates a synchronization of responses across different items, indicating a high level of accountability and honesty among the survey participants. The emphasis on novel and creative approaches reflects their thoughtful reflections and documented managerial experiences.

Improving the quality of both initial and ongoing training ranks fourth overall, as indicated by 16.39% of respondents. This finding suggests that respondents may have undervalued the importance of training in maintaining competitiveness in the educational market. The effectiveness of the didactic process and subsequent school performance significantly depend on the expertise of the trainers, making quality training a crucial aspect of remaining competitive.

On the same level, there is a notable emphasis on the development of a quality culture, identified by 9.83% of respondents. Given that Total Quality Management (TQM) emphasizes full member participation, long-term success, client satisfaction, and organizational benefits, prioritizing the development of a quality culture is crucial in the current sociocultural context, the findings highlight the importance of increased institutional autonomy, educational marketing strategies, quality training, and the development of a quality culture as key actions for colleges to remain competitive. By embracing these directions, college managers can navigate the challenges of the educational market, adapt to changing dynamics, and position their institutions for long-term success.

The curriculum implemented in schools serves as a tangible expression of institutional autonomy, and its effectiveness relies on its ability to address the actual needs of students. To delve deeper into the relationship between the school-based curriculum and students' requirements, we adopted an innovative approach to educational administration, incorporating a specific question in the questionnaire.

The results unveiled a significant proportion of respondents (21.31% very much and 45.905% much) who perceived a strong alignment between the school-based curriculum and the actual needs of the students. This notable percentage suggests a gratifying outcome, indicating that the curriculum has been thoughtfully designed to cater to the diverse needs of the student population. The positive correlation between the curriculum and students' actual requirements showcases the educational institution's responsiveness to the ever-evolving and dynamic needs of the learners.

Nevertheless, it is important to acknowledge that a smaller segment of respondents (24.60%) acknowledged only a minor correspondence between the school-based curriculum and the actual requirements of the children. This finding underscores the need for improvement in ensuring a more precise alignment between the curriculum and the specific needs of students. Addressing this discrepancy presents an opportunity to enhance the curriculum's effectiveness and create a more tailored learning experience that aligns with the individualized learning needs, interests, and aspirations of students.

By employing this innovative approach to educational administration, this study provides valuable insights into how the curriculum is perceived in terms of its responsiveness to students' actual requirements. The findings underscore the significance of ongoing assessment and adaptation of the curriculum to ensure its relevance, engagement, and impact for all students. By continuously aligning the curriculum with the evolving needs and aspirations of the students, educational institutions can establish an inclusive and effective learning environment that fosters student success and growth as mentioned in TABLE 9.5..

The level of institutional autonomy and the effectiveness of educational management can be reflected in the manager's influence and decision-making authority over the selection of teaching staff in the school unit. The questionnaire included a question specifically addressing the involvement of the manager in the selection process, and the results are presented in Table 9.6.

TABLE 9.5
Correlation between the School-Based Curriculum and the Students' Real Needs

Item	Very high	High	Low	None at all	Total
State the extent to which the school-based curriculum corresponds with the students' real needs	13	28	15	%	61
	21.31%	45.90%	24.60%	8.02%	100%

TABLE 9.6
The Need for the Manager to Get Involved in Hiring Teaching Staff

Item	N	%
Yes	41	67.21%
No	12	19.67%
I cannot tell	8	13.11%
Total	61	100%

A significant percentage of managers (67.21%) who participated in the survey expressed their support and favorability towards the manager's engagement in the selection and employment of teaching staff at the school unit level. This indicates a recognition of the manager's role in shaping the teaching staff and ensuring the quality of education within the institution. It also highlights the importance of managerial expertise and involvement in personnel decisions, including hiring and recruitment.

On the other hand, a notable proportion (19.63%) of managers held a different perspective on this matter. This divergence of opinion may be attributed to various factors. Some managers may acknowledge the diverse responsibilities and activities involved in educational management and prefer to delegate certain aspects of the selection process to other stakeholders. This could stem from a desire to share the responsibility or to benefit from the expertise of other individuals involved in the recruitment process.

Furthermore, a smaller subset of managers (13.11%) did not provide a clear response to this question, indicating a lower level of accountability and decision-making autonomy within institutional management. This observation suggests that there may be room for improvement in terms of fostering a greater sense of responsibility and self-accountability among managers in the selection and employment of teaching staff.

The differing perspectives and responses on the manager's involvement in the selection of teaching staff highlight the complex nature of educational management and the varying approaches adopted by different individuals. Balancing the need for managerial control and decision-making with the benefits of collaboration and shared responsibility is a challenge that educational institutions must address.

Overall, the findings underscore the significance of the manager's role in the selection of teaching staff and the need to establish clear guidelines and procedures to ensure transparency, fairness, and quality in the recruitment process. Promoting a greater sense of responsibility and decisional self-accountability among managers can contribute to effective educational management and the enhancement of teaching staff quality in the school unit.

9.5 LIMITATIONS

It is essential to acknowledge the potential limitations of this research. Firstly, the study's findings may be context-specific and may not be generalizable to all higher education institutions. Secondly, self-reporting bias may exist in survey responses, and social desirability bias could influence participant perspectives. To mitigate these limitations, a diverse sample from various institutions will be targeted, and triangulation of data sources will be employed to enhance the validity and reliability of the findings.

9.6 SIGNIFICANCE

This research methodology will provide a comprehensive understanding of the impact of NEP-compliant curriculum changes on LMS effectiveness in higher education institutions. By combining quantitative and qualitative data, the study will capture both numerical trends and the lived experiences of stakeholders. The findings will contribute to evidence-based decision-making, inform institutional strategies for curriculum reform and LMS implementation, and guide future research and development efforts in the field of educational technology.

9.7 CONCLUSION

In conclusion, as HEIs strive to adapt to the changing educational landscape, it is essential to explore avenues for enhancing the effectiveness of LMS platforms. By aligning the curriculum with NEP guidelines, institutions can potentially unlock the benefits of holistic and multidisciplinary learning experiences, critical thinking skills, and increased student engagement. This study aims to investigate the impact of NEP-compliant curriculum changes on LMS effectiveness, providing valuable insights for educational stakeholders seeking to leverage policy-driven modifications to improve their educational systems. Tis study's significance lies in its potential to contribute to educational reform, enhance learning outcomes, improve administrative efficiency, support faculty adoption, inform institutional decision-making, stimulate future research and development, promote accessibility and inclusivity, foster lifelong learning and skills development, cater to the needs of digital natives, and address technological challenges. By addressing these areas, the study can support the broader objective of improving the quality of higher education and aligning it with the goals of the National Education Policy. Understanding the impact of NEP-compliant curriculum changes on LMS effectiveness is crucial for institutions aiming to align their educational practices with the goals of the NEP and maximize the potential of technology-enhanced learning environments. Further research is needed to explore the combined effect of these factors and address the challenges and opportunities that arise during implementation.

BIBLIOGRAPHY

Aktar, S. "New education policy 2020 of India: A theoretical analysis." *International Journal of Business and Management Research*, vol. 9, no.3, pp. 302–306, 2021.

Aytac, A. and Deniz, V. "Quality function deployment in education: A curriculum review." *Quality and Quantity*, vol. 39, pp. 507–514, August 2005.

Bechard, J. P. "L'enseignement superieur et les innovations pedagogiques: une recension des ecrits." *Revue des sciences de l'*, vol. 27, no. 2, pp. 257–281.

Birkinshaw, J., Hamel, G. and Mol, M. "Management innovation." Academy of Management Review, vol. 33, no. 4, pp. 825–845, 2008.

Bunăiaşu, C. M. and Strungă, A.-C. "A potential methodological tool in order to plan the curriculum in school." *Procedia - Social and Behavioral Sciences*, vol. 76, pp. 140–145, 2013.

Ciumara, T. "Explorarea functiilor economice ale consultantei in practicile innovative de management in conditiile dezvoltarii economice durabile." Institutul National De Cercetari Economice, 2013.

Crişan, A. and Enache, R. "Designing customer-oriented courses and curricula in higher education. A possible model." *Procedia - Social and Behavioral Sciences*, vol. 11, pp. 235–239, 2011.

Deng, Z. "Curriculum planning and systems change." *International Encyclopedia of Education (Third Edition)*, B. McGaw, E. Baker, and P. Penelope, Eds. Oxford: Elsevier, 2010, pp. 384–389.

Denton, J. W., Franke, V. and Surendra, K. N. "Curriculum and course design: A new approach using quality function deployment." *Journal of Education for Business*, vol. 81, pp. 111–117, November 2005.

Dressel, P. L. *Improving Degree Programs: A Guide to Curriculum Development, Administration, and Review*. San Francisco: Jossey-Bass, 1980.

Foreman, J. et al. "Association between digital smart device use and myopia: a systematic review and meta-analysis." *The Lancet Digital Health*, vol. 3, no. 12, pp. e806–e818, 2021.

Gonzalez, M. E., Quesada, G., Mueller, J., and Mueller, R. D. "International business curriculum design: Identifying the voice of the customer using QFD." *Journal of International Education in Business*, vol. 4, pp. 6–29, 2011.

Graham, Charles R. "Exploring definitions, models, frameworks, and theory for blended learning research." In A. G. Picciano, C. D. Dziuban, C. R. Graham, and P. D. Moskal, Eds. *Blended Learning*. Routledge, 2021, pp. 10–29.

Halliburton, D. "Designing curriculum." In *Developing the College Curriculum*, A. W. Chickering, D. Halliburton, W. H. Berquist, and J. Lindquist, Eds. Washington, DC: Council for the Advancement of Small colleges, 1977, pp. 51–74.

Hamel, G. and Breen, B. *Viitorul managementului*. Publica Publishing House, Bucharest, 2010.

Ho, S. S. et al. "Understanding public willingness to pay more for plant-based meat: Environmental and health consciousness as precursors to the influence of presumed media influence model." *Environmental communication*, vol. 16, no. 4, pp. 520–534, 2022.

Hussain, S. et al. "Role of micronutrients in salt stress tolerance to plants." In *Plant Nutrients and Abiotic Stress Tolerance*, M. Hasanuzzaman, M. Fujita, H. Oku, K. Nahar, and B. Hawrylak-Nowak, Eds. Springer, 2018, pp. 363–376.

Iosifescu, S. et al. "Management institutional si management de project." Educatia 2000+, Bucuresti, 2009.

Ng, Y. X. et al. "Assessing mentoring: A scoping review of mentoring assessment tools in internal medicine between 1990 and 2019." *PloS One* vol. 15, no. 5, p. e0232511, 2020.

Pérez, J. S. and Aleu, F. G. "Industrial engineering approach to develop an industrial engineering curriculum." IIE Annual Conference. Proceedings. Institute of Industrial and Systems Engineers, 2009, pp. 271–277.

White, H. D. "Computing a curriculum: Descriptor-based domain analysis for educators." *Information Processing & Management*, vol. 37, pp. 91–117, 2001.

Yousafzai, S. et al. "Spatio-temporal assessment of land use dynamics and urbanization: linking with environmental aspects and DPSIR framework approach." *Environmental Science and Pollution Research*, vol. 29, no. 54, pp. 81337–81350, 2022.

Yang Yang, De-Chuan Zhan, and Yuan Jiang. "Learning by actively querying strong modal features". In *Proceedings of the 25th International Joint Conference on Artificial Intelligence*, 2016, pp. 2280–2286. New York, NY.

Additive Manufacturing

10 The Future of Immersive Experience

Exploring Metaverse Application Development Technologies and Tools

*V. Sheeja Kumari, S. Ponmaniraj,
G. Vennira Selvi, and Simarjeet Kaur*

10.1 INTRODUCTION

In recent years, the concept of the Metaverse has gained widespread attention and excitement, thanks in part to the success of virtual worlds like Second Life and the rise of technologies like virtual reality and blockchain. The Metaverse is envisioned as a fully immersive virtual universe where users can interact with each other and with digital objects and environments in ways that are not possible in the physical world [1].

As the Metaverse continues to evolve, developers are working to create new applications and experiences that will enable users to explore this exciting new world. The potential applications of the Metaverse are numerous, from gaming and entertainment to education, social networking, and even e-commerce.

In this chapter, we will explore the technologies and tools that are driving the development of the Metaverse. We will examine the underlying technologies, such as blockchain, virtual reality, augmented reality, and artificial intelligence, and how they are being used to create immersive experiences. We will also explore the tools and platforms that are available for Metaverse application development, and the challenges that developers face in creating these experiences. By gaining a better understanding of the technologies and tools that are shaping the Metaverse, developers can create engaging and innovative experiences for users in this emerging field. We hope that this paper will serve as a useful resource for developers, researchers, and anyone interested in the future of immersive experiences [2].

The proposed system for Metaverse application development aims to create a fully immersive and decentralized virtual world for users. The system integrates various technologies and tools, such as game engines, virtual reality, augmented reality, blockchain, decentralized platforms, and artificial intelligence, to create engaging and innovative experiences for users. To create 3D environments and objects, a game engine like Unity or Unreal could be used. The game engine could also be integrated with virtual reality and augmented reality technologies to provide users with an immersive experience within the Metaverse.

To enable user ownership of virtual assets and facilitate transactions and interactions between users, blockchain technology and smart contracts could be used [3]. The system could also incorporate a decentralized platform like Decentraland or Somnium Space to create interconnected virtual worlds.

To ensure that the Metaverse operates in a fair and responsible manner, the system could incorporate governance mechanisms like decentralized autonomous organizations (DAOs). This would allow users to participate in the decision-making process and ensure that the Metaverse operates in a decentralized and democratic manner [4]. Finally, the system could utilize artificial intelligence to create intelligent NPCs (non-playing characters) that respond to user actions and provide a more realistic and personalized experience for users. This would enhance the overall immersion and engagement within the Metaverse.

10.2 LITERATURE SURVEY

The concept of the Metaverse has been explored in various forms of literature, including science fiction, philosophy, and even academic research. In recent years, as technology has advanced, the concept has become more of a reality, and many researchers and industry experts have explored its potential applications and implications.

One important work in this area is Neal Stephenson's 1992 novel *Snow Crash*, which is credited with popularizing the term "Metaverse" and introducing many of the key concepts that are now associated with the idea. In the novel, the Metaverse is a fully immersive virtual world where users can interact with each other and with digital objects and environments.

In academic research, scholars have explored the potential applications of the Metaverse in various fields, such as education, healthcare, and business. For example, in a 2018 article published in the *Journal of Educational Technology & Society*, researchers explored the potential of the Metaverse as a platform for immersive learning environments.

In terms of technology, blockchain has emerged as a key enabling technology for the Metaverse, allowing for decentralized ownership of digital assets and environments. In a 2020 article published in the IEEE Computer Society's Computer magazine, researchers explored the use of blockchain for virtual worlds, highlighting its potential for creating a more equitable and decentralized Metaverse.

Virtual reality and augmented reality technologies are also key components of the Metaverse, enabling users to immerse themselves in digital environments and interact with them in ways that are not possible in the physical world. In a 2019 article published in the International Journal of Human-Computer Interaction, researchers explored the use of VR and AR in the context of the Metaverse, highlighting their potential for creating immersive and engaging experiences.

In recent years, there has been a surge of interest in the Metaverse from both industry and academia. Companies such as Facebook and Epic Games are investing heavily in Metaverse development, and research institutions are exploring the potential applications of the Metaverse in fields such as education and healthcare.

One recent study by Tsekleves et al. (2021) explored the potential of the Metaverse to support mental health and wellbeing. The authors argued that immersive technologies like the Metaverse could be used to create engaging and therapeutic experiences for users, and could potentially have important applications in the treatment of mental health conditions.

Another study by Wang et al. (2021) explored the use of blockchain technology in the Metaverse. The authors argued that blockchain could be used to create a decentralized and interoperable Metaverse, which would enable users to move seamlessly between different virtual worlds.

1. "The Metaverse: A Vision for a Virtual 3D Social Network" by Julian Dibbell. In this article, Dibbell explores the concept of the Metaverse as a virtual 3D social network and examines its potential applications in various industries. He discusses the underlying technologies that enable the Metaverse, such as virtual reality, blockchain, and AI.

2. "The Potential of Blockchain in Virtual Reality" by Martin Coggins. This paper explores how blockchain can be used to create decentralized virtual worlds and enable user ownership of virtual assets. The paper also discusses the potential for blockchain to address issues of scalability and interoperability in the Metaverse.

3. "Unity vs. Unreal: How to Choose the Best Game Engine" by Daniel do Nascimento. This article compares the two most popular game engines, Unity and Unreal, and discusses their suitability for Metaverse application development. The article also provides an overview of other game engines and development tools that can be used for Metaverse development.

4. "Decentralized Autonomous Organizations and the Future of Governance" by Primavera De Filippi and Aaron Wright. This paper discusses the potential for decentralized autonomous organizations (DAOs) to address governance issues in the Metaverse. The paper examines the use of DAOs in virtual worlds like Second Life and explores their potential for use in the Metaverse.

5. "The Future of the Metaverse: A Roadmap" by Jamey Harvey. This paper provides a roadmap for the development of the Metaverse and examines the challenges that developers face in creating this immersive virtual world. The paper discusses issues of governance, interoperability, and scalability and proposes solutions for addressing these challenges.

6. *The Metaverse: A Roadmap for the Future* by Don Tapscott and Alex Tapscott (2019). This book provides a comprehensive overview of the potential of the Metaverse and how it could impact various industries, including entertainment, education, and healthcare. The authors explore the underlying technologies, such as blockchain and virtual reality, and discuss how they could be used to create immersive experiences in the Metaverse.

7. *Decentralized Applications: Harnessing Bitcoin's Blockchain Technology* by Siraj Raval (2018). In this book, the author explores how blockchain technology can be used to create decentralized applications (dApps) and Metaverse experiences. The book covers topics such as smart contracts, decentralized

storage, and decentralized identity, and discusses how they can be used in Metaverse application development

8. *Virtual Reality and Augmented Reality: Myths and Realities* by Philippe Fuchs (2019). This book provides an overview of the current state of virtual reality and augmented reality technologies and how they can be used to create immersive experiences. The author discusses the challenges that developers face in creating compelling experiences and explores the potential of these technologies in the context of the Metaverse.

9. *Building Virtual Reality with Unity and Steam VR* by Jeff W. Murray (2019). This book provides a hands-on guide to creating virtual reality experiences using the Unity game engine and Steam VR. The author covers topics such as locomotion, interaction, and user interfaces, and discusses how these concepts can be applied in the context of the Metaverse.

10. *Blockchain Basics: A Non-Technical Introduction in 25 Steps* by Daniel Drescher (2017). In this book, the author provides a beginner-friendly introduction to blockchain technology and its potential applications. The book covers topics such as consensus algorithms, public and private blockchains, and smart contracts, and discusses how these concepts can be used in the context of Metaverse application development.

By gaining a better understanding of the technologies and tools that are shaping the Metaverse, developers can create engaging and innovative experiences for users in this emerging field. We hope that this paper will serve as a useful resource for developers, researchers, and anyone interested in the future of immersive experiences. Overall, these papers provide a comprehensive overview of the technologies and tools that are shaping the future of Metaverse application development. They discuss the potential applications of the Metaverse in various industries, the underlying technologies that enable it, and the challenges that developers face in creating immersive experiences for users.

10.3   METAVERSE ARCHITECTURE

The term "metaverse" refers to a virtual reality (VR) or augmented reality (AR) space where users can interact with a computer-generated environment and other users in real time. It is often described as a collective virtual shared space that encompasses the physical world and digital realms. The concept of the metaverse has gained significant attention in recent years due to advancements in technology and the increasing popularity of immersive experiences. It is envisioned as a fully realized digital universe where people can engage in various activities, such as socializing, working, gaming, exploring, and conducting business.

In the metaverse, users can create digital representations of themselves, known as avatars, to navigate and interact with the virtual environment. They can communicate with other users through voice chat, text, or gestures, creating a sense of presence and shared experience [5].

Companies like Facebook (now Meta), Microsoft, and Epic Games have expressed their intentions to develop metaverse platforms or ecosystems. These

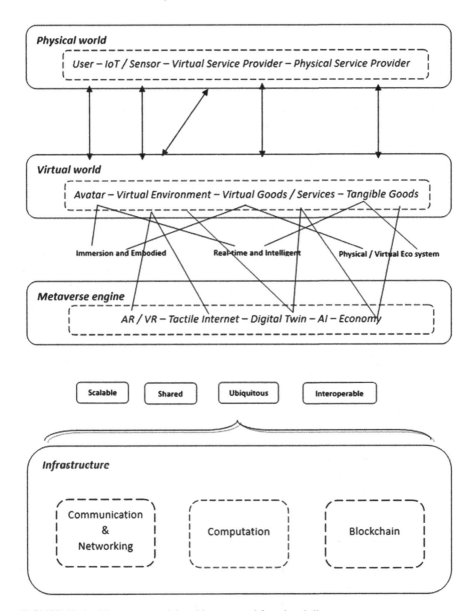

FIGURE 10.1   Metaverse model architecture and functional diagram.

platforms aim to provide a seamless and interconnected virtual space that integrates various services and experiences, including virtual commerce, entertainment, education, and more.

It is important to note that the concept of the metaverse is still evolving, and its ultimate form and scope are yet to be fully realized. The term is often used broadly to describe a vision of an immersive, interconnected digital space, but there are still technological, social, and ethical challenges to overcome before a true metaverse becomes a reality.

The architecture of the metaverse refers to the underlying infrastructure and design principles that enable the creation and functioning of a virtual reality (VR) or augmented reality (AR) space where users can interact with each other and the virtual environment as shown in FIGURE 10.1. While there is no universally defined architecture for the metaverse yet, several key components and concepts are often discussed in relation to its design. Here are some aspects commonly associated with metaverse architecture:

1. *Virtual worlds:* The metaverse consists of interconnected virtual worlds or environments. These worlds can vary in scale, design, and purpose. They may be created by individuals, organizations, or game developers and can encompass everything from social spaces to gaming realms, educational simulations, virtual marketplaces, and more.

2. *Avatar system:* In the metaverse, users typically create avatars, which are digital representations of themselves. Avatars enable users to navigate and interact within the virtual environment. They can customize their avatars' appearance, clothing, and accessories to express their individuality.

3. *Networking infrastructure:* The metaverse relies on robust networking infrastructure to connect users and facilitate real-time interactions. This involves the use of servers, cloud computing, and high-speed internet connections to ensure seamless communication and synchronization across virtual worlds.

4. *Interoperability and standards:* To achieve a cohesive metaverse experience, interoperability and standards are crucial. This means that different platforms, applications, and virtual worlds should be able to communicate and share data seamlessly. Open standards and protocols facilitate interoperability, allowing users to move their avatars and assets between different metaverse ecosystems.

5. *Spatial computing:* Spatial computing technologies play a significant role in the metaverse. They enable the seamless integration of virtual content with the physical world, allowing users to interact with virtual objects and experiences overlaid onto their real-world environment. This can be achieved through technologies like augmented reality (AR), mixed reality (MR), or advanced VR systems.

6. *Decentralization and blockchain:* Some envision the metaverse as a decentralized system using blockchain technology. Blockchain can provide transparency, security, and ownership verification for virtual assets, including virtual currencies, digital goods, and collectibles. This can empower users with greater control and ownership of their digital identities and assets within the metaverse.

7. *Artificial Intelligence (AI):* AI plays a role in the metaverse by enabling intelligent NPCs that can interact with users, create dynamic environments, and enhance the overall user experience. AI algorithms can also help personalize content, recommend activities, and assist with various tasks within the metaverse.

It is important to note that the architectural aspects of the metaverse are still being explored, and there is ongoing research and development in this area. The ultimate architecture of the metaverse will likely be shaped by technological advancements, user needs, and the collaborative efforts of various stakeholders in the virtual reality and augmented reality industries.

10.4 TECHNOLOGIES AND TOOLS FOR METAVERSE APPLICATION DEVELOPMENT

The development of Metaverse applications involves the integration of several technologies and tools to create a fully immersive and engaging virtual world [3]. Some of the key technologies and tools used in Metaverse application development are:

1. *Game engines:* Game engines like Unity and Unreal are widely used in Metaverse application development to create 3D environments and objects. These engines provide tools for designing, modeling, and programming virtual objects and environments.
2. *Virtual reality and augmented reality technologies:* Virtual reality and augmented reality technologies are used to provide an immersive experience for users. Virtual reality technologies create a completely artificial environment, while augmented reality technologies overlay virtual objects onto the real world.
3. *Blockchain technology and smart contracts:* Blockchain technology is used to create a secure and decentralized system for transactions and interactions within the Metaverse. Smart contracts can be used to automate transactions and interactions between users.
4. *Decentralized platforms:* Decentralized platforms like Decentraland and Somnium Space allow for the creation of interconnected virtual worlds. These platforms provide users with a shared space to interact and transact within the Metaverse.
5. *Artificial Intelligence:* Artificial intelligence can be used to create intelligent NPCs that provide dynamic and personalized interactions for users. AI can also be used for data analysis and modeling to improve the overall user experience.

The integration of these technologies and tools in Metaverse application development can provide users with a fully immersive and engaging virtual world [3,4]. They allow for user ownership of virtual assets, facilitate transactions and interactions between users, and provide a fair and responsible governance mechanism.

10.4.1 GAME ENGINES (UNITY, UNREAL)

Game engines like Unity and Unreal are widely used in Metaverse application development to create 3D environments and objects. These engines provide tools for designing, modeling, and programming virtual objects and environments [3].

Unity is a popular game engine used in Metaverse application development due to its ease of use and flexibility. It provides tools for creating 2D and 3D environments, as well as support for virtual reality and augmented reality technologies. Unity also has a large community of developers, which provides a wealth of resources and support for Metaverse application development.

Unreal is another popular game engine used in Metaverse application development. It provides advanced tools for creating high-quality 3D environments and objects, as well as support for virtual reality and augmented reality technologies. Unreal also provides advanced programming capabilities, making it suitable for complex Metaverse applications.

Both Unity and Unreal provide support for cross-platform development, allowing Metaverse applications to be developed for multiple platforms including desktop, mobile, and console. They also provide support for scripting languages like C# and C++, making it easier to program complex Metaverse applications. The game engines like Unity and Unreal are essential tools in Metaverse application development [4]. They provide the necessary tools for creating 3D environments and objects, as well as support for virtual reality and augmented reality technologies. Additionally, their cross-platform support and programming capabilities make them suitable for complex Metaverse applications.

10.4.2 Virtual Reality and Augmented Reality Technologies

Virtual reality (VR) and augmented reality (AR) technologies are essential in Metaverse application development as they provide users with an immersive experience [6]. Virtual reality technologies create a completely artificial environment, which can be explored by users through a VR headset or other VR devices. This allows users to interact with the virtual environment and other users in a highly immersive way. Virtual reality technologies can be used to create a range of Metaverse applications, including virtual social spaces, educational environments, and gaming applications.

Augmented reality technologies, on the other hand, overlay virtual objects onto the real world. This can be achieved through devices like smartphones, tablets, and AR glasses. Augmented reality technologies can be used in Metaverse applications to create interactive experiences in real-world environments. For example, AR technologies can be used to create interactive museum exhibits, or to provide users with virtual guides for exploring real-world locations.

Both VR and AR technologies are rapidly advancing, with new hardware and software solutions being developed regularly. This is leading to an increasingly realistic and immersive experience for users. Additionally, the development of WebVR and WebAR technologies is making it easier to develop Metaverse applications that can be accessed directly through a web browser, without the need for specialized hardware or software [7].

10.4.3 Blockchain Technology and Smart Contracts

Blockchain technology and smart contracts are essential in Metaverse application development, as they provide a secure and decentralized system for transactions and interactions within the virtual world [8]. Blockchain technology provides a decentralized system for storing and verifying data, which makes it ideal for creating a transparent and secure system for transactions within the Metaverse. Blockchain technology enables users to own and trade virtual assets, such as in-game items and digital collectibles, in a secure and transparent manner. This technology also enables users to have control over their personal data and protects them from potential fraud or theft.

Smart contracts, which are self-executing computer programs that automatically enforce the terms of a contract, can be used to automate transactions and interactions between users within the Metaverse. Smart contracts can be used to enable the exchange of virtual assets, such as digital currency and virtual goods, without the need for intermediaries.

The integration of blockchain technology and smart contracts in Metaverse application development can also provide a fair and responsible governance mechanism. This can help ensure that decisions within the Metaverse are made democratically and in the best interests of the community as a whole.

10.4.4 DECENTRALIZED PLATFORMS (DECENTRALAND, SOMNIUM SPACE)

Decentralized platforms like Decentraland and Somnium Space are becoming increasingly popular in Metaverse application development due to their decentralized nature and ability to provide users with greater control over their virtual environments and experiences.

Decentraland is a decentralized virtual world built on the Ethereum blockchain. Users can create, experience, and monetize content and applications within the virtual world. The ownership and control of land and virtual assets within Decentraland are stored on the blockchain, giving users complete control over their virtual assets. This also means that virtual assets can be traded and exchanged on open markets, creating a unique virtual economy within the Metaverse [5].

Somnium Space is another decentralized virtual world that uses blockchain technology to provide users with greater control over their virtual experiences. Users can create, own, and monetize content and applications within the virtual world. The ownership and control of land and virtual assets within Somnium Space are also stored on the blockchain, providing users with complete control over their virtual assets [5,8].

Both Decentraland and Somnium Space enable users to create and experience virtual environments that are truly decentralized, allowing users to control their virtual assets and experiences. This creates a unique virtual world where users can explore, create, and interact in a way that is not possible in traditional centralized systems.

10.5   ARTIFICIAL INTELLIGENCE

Artificial Intelligence (AI) is becoming increasingly important in Metaverse application development as it can enable more realistic and immersive experiences for users within the virtual world. One of the most significant applications of AI in the Metaverse is in creating realistic virtual characters and NPCs. AI-powered virtual characters can be programmed to interact with users in a way that feels more natural and human-like. This can create a more immersive and engaging experience for users within the virtual world [9]. AI can also be used to enhance the graphics and visuals within the virtual world. AI-powered algorithms can be used to create more realistic and detailed environments, characters, and objects within the virtual world. This can provide users with a more immersive and visually stunning experience.

AI-powered chatbots can also be used within the Metaverse to provide users with personalized and interactive experiences. Chatbots can be programmed to provide information, assistance, and support to users within the virtual world. This can create a more engaging and personalized experience for users within the Metaverse.

10.6 FUTURE SYSTEM FOR METAVERSE APPLICATION DEVELOPMENT

A future system for Metaverse application development could involve the integration of various technologies and tools to create immersive experiences for users. The system could involve the use of a game engine, such as Unity or Unreal, to create 3D environments and objects. The game engine could be integrated with virtual reality and augmented reality technologies to enable users to interact with the environment and objects in a more immersive way [10].

In addition, the system could incorporate blockchain technology to enable user ownership of virtual assets and the creation of decentralized virtual worlds. Smart contracts could be used to facilitate transactions and interactions between users within the Metaverse.

To address issues of scalability and interoperability, the system could utilize a decentralized platform, such as Decentraland or Somnium Space, that allows for the creation of interconnected virtual worlds. The platform could also incorporate governance mechanisms, such as decentralized autonomous organizations (DAOs), to ensure that the Metaverse operates in a fair and responsible manner [11]. To create engaging and innovative experiences for users, the system could incorporate artificial intelligence to enable dynamic and personalized interactions within the Metaverse. AI could be used to create intelligent NPCs that respond to user actions and provide a more realistic and immersive experience.

The future system for Metaverse application development would integrate various technologies and tools to create a fully immersive and decentralized virtual world. The system would enable user ownership of virtual assets, facilitate transactions and interactions between users, and provide a fair and responsible governance mechanism. With the incorporation of AI, the system would also provide dynamic and personalized interactions for users.

10.6.1 INTEGRATION OF TECHNOLOGIES AND TOOLS FOR A FULLY IMMERSIVE AND DECENTRALIZED VIRTUAL WORLD

The integration of different technologies and tools is crucial in creating a fully immersive and decentralized virtual world within the Metaverse. By combining various technologies and tools, developers can create a more realistic, interactive, and engaging virtual world for users [12]. For example, integrating game engines like

Unity or Unreal with virtual reality and augmented reality technologies can create a more immersive experience for users within the virtual world. Users can interact with virtual environments in a more natural and intuitive way, creating a more realistic and engaging experience.

The integration of blockchain technology and smart contracts can also provide users with greater control over their virtual assets and experiences within the Metaverse. By using blockchain technology, ownership and control of virtual assets can be stored securely and transparently, providing users with complete control over their assets. Smart contracts can also be used to automate transactions and interactions within the virtual world, creating a more seamless and efficient experience [13]. Artificial intelligence can also be integrated into the virtual world to create more realistic and interactive experiences. AI-powered virtual characters and NPCs can provide users with a more natural and human-like interaction, while AI-powered chatbots can provide users with personalized assistance and support within the virtual world.

10.6.2 Use of Game Engines to Create 3D Environments and Objects

Game engines like Unity and Unreal are widely used in Metaverse application development to create 3D environments and objects. These engines provide developers with a suite of tools and features to create and manipulate virtual objects and environments within the Metaverse.

Using game engines, developers can create 3D models of virtual environments, including landscapes, buildings, and objects, and add interactive elements like physics simulations, lighting effects, and animations. These models can then be imported into the Metaverse, where users can interact with them in real time [14]. Game engines also provide developers with a range of scripting tools and programming languages, such as C# and C++, to create interactive gameplay mechanics and logic within the Metaverse. This allows developers to create complex interactive experiences for users, such as puzzles, quests, and multiplayer games.

Another benefit of using game engines in Metaverse application development is that they allow for cross-platform development. Game engines can be used to create applications that can run on a range of platforms, including desktops, mobile devices, and virtual and augmented reality devices.

10.6.3 Integration of Virtual Reality and Augmented Reality Technologies for an Immersive Experience

Virtual reality (VR) and augmented reality (AR) technologies are increasingly being integrated into Metaverse application development to create a more immersive experience for users. VR and AR technologies allow users to interact with virtual environments in a more natural and intuitive way, providing a more immersive and engaging experience. Virtual reality technology allows users to enter a fully immersive virtual environment, where they can interact with objects and characters in a more natural and intuitive way. VR headsets and controllers enable users to move around and interact with virtual objects in a more realistic way, creating a more immersive experience.

In the Metaverse, VR technology can be used to create virtual environments where users can explore, interact with objects, and engage in social interactions with other users [15].

Augmented reality technology, on the other hand, allows users to overlay digital content onto the real world. AR technology is commonly used in mobile devices, where users can point their camera at an object and see digital information overlaid onto it. In the Metaverse, AR technology can be used to enhance the user's experience by overlaying digital information onto virtual objects or environments, providing a more immersive and interactive experience.

The integration of VR and AR technology in Metaverse application development can also provide new opportunities for social interaction. For example, users can use VR headsets to enter a virtual world where they can meet and interact with other users in a more natural and intuitive way, creating a more immersive and engaging social experience.

10.6.4   USE OF BLOCKCHAIN TECHNOLOGY AND SMART CONTRACTS FOR USER OWNERSHIP OF VIRTUAL ASSETS AND TRANSACTIONS

Blockchain technology and smart contracts are increasingly being used in Metaverse application development to enable user ownership of virtual assets and transactions. Blockchain technology is a decentralized ledger system that provides a secure and transparent way to store and transfer data. Smart contracts are self-executing programs that run on blockchain networks and can automate the transfer of assets and other transactions [16]. In the context of the Metaverse, blockchain technology and smart contracts can be used to enable user ownership of virtual assets such as virtual real estate, items, and even virtual currencies. Using blockchain technology, users can have secure and transparent ownership of their virtual assets, and can transfer them to other users in a peer-to-peer manner, without the need for intermediaries.

Smart contracts can automate the transfer of virtual assets and transactions, enabling users to conduct transactions in a more efficient and transparent manner. For example, when a user purchases virtual real estate within the Metaverse, a smart contract can automatically transfer ownership of the asset to the user, without the need for intermediaries. Similarly, when a user sells a virtual asset, a smart contract can automatically transfer ownership to the buyer and transfer payment to the seller. The use of blockchain technology and smart contracts in Metaverse application development can also provide new opportunities for monetization [15,16]. For example, users can earn virtual currencies for completing tasks or contributing to the Metaverse community, and these currencies can be exchanged for real-world currencies or other virtual assets.

10.6.5   INCORPORATION OF DECENTRALIZED PLATFORMS FOR INTERCONNECTED VIRTUAL WORLDS

Decentralized platforms are becoming increasingly important in Metaverse application development, as they enable the creation of interconnected virtual worlds. Decentralized platforms allow for the creation of virtual environments that are owned and controlled by the users themselves, rather than by a centralized entity. This can enable greater autonomy and creativity in the development of virtual environments, as well as greater opportunities for user-driven content creation [15].

Decentralized platforms such as Decentraland and Somnium Space enable the creation of interconnected virtual worlds by using blockchain technology and smart contracts to enable ownership and control of virtual land and assets. Users can purchase virtual land within these platforms and build their own virtual environments, which can be connected to other users' environments to create a larger interconnected world.

This interconnected world can enable new opportunities for social interaction, as users can explore and interact with other users' virtual environments. It can also provide new opportunities for content creation, as users can collaborate to create larger and more complex virtual environments. The use of decentralized platforms also enables greater user control over their virtual environments, as they are not subject to the control of a centralized entity. This can enable greater creativity and experimentation in the development of virtual environments, as well as greater opportunities for user-driven innovation.

10.6.6 Use of Decentralized Autonomous Organizations (DAOs) for FAIR and Responsible Governance

Decentralized autonomous organizations (DAOs) are another important technology in Metaverse application development, as they enable fair and responsible governance within the Metaverse. DAOs are decentralized organizations that are governed by smart contracts and run on blockchain networks. They enable transparent and democratic decision-making processes, as all members have equal voting rights and can participate in the decision-making process [16].

In the context of the Metaverse, DAOs can be used to govern virtual worlds and assets, ensuring that they are managed in a fair and responsible manner. For example, a DAO could be created to govern a particular virtual world, with members of the DAO having equal voting rights in decisions related to the management of the world. This can help to ensure that the world is managed in a transparent and democratic manner, with decisions being made in the best interests of the community as a whole. DAOs can also be used to manage virtual assets within the Metaverse, ensuring that they are distributed fairly and in a way that benefits the community as a whole. For example, a DAO could be created to manage the distribution of a particular virtual currency within the Metaverse, with members of the DAO having equal voting rights in decisions related to the distribution of the currency. The use of DAOs in Metaverse application development can help to ensure that virtual worlds and assets are managed in a fair and responsible manner, with decisions being made in the best interests of the community as a whole. This can help to promote greater trust and collaboration within the Metaverse ecosystem, enabling the development of a more vibrant and sustainable virtual world [17].

10.6.7 Utilization of Artificial Intelligence to Create Intelligent NPCs for a More Realistic and Personalized Experience

Artificial intelligence (AI) is an important technology in Metaverse application development, as it can be used to create intelligent non-player characters (NPCs) that provide a more realistic and personalized experience for users. NPCs are characters within virtual worlds that are controlled by the computer rather than by a human player. They are an important component of many virtual worlds, as they provide additional challenges and opportunities for social interaction. By using AI, developers can create NPCs that are more intelligent and responsive to user actions [16]. For example, AI can be used to create NPCs that learn from user behavior and adapt their behavior accordingly, providing a more personalized experience for the user. AI can also be used to create NPCs that are more realistic in their interactions, providing a more immersive experience for the user.

One of the main benefits of using AI to create intelligent NPCs is that it enables greater user engagement within the Metaverse. Users are more likely to be engaged with virtual worlds that feature intelligent NPCs that provide a more personalized and immersive experience. This can lead to greater user retention and a more vibrant and sustainable virtual world ecosystem. In addition to creating intelligent NPCs, AI can also be used to optimize other aspects of virtual world development [17]. For example, AI can be used to optimize virtual world physics, ensuring that objects within the virtual world behave in a realistic manner. AI can also be used to optimize virtual world graphics, ensuring that the virtual world is visually appealing and responsive to user actions.

10.7 METAVERSE SECURITY CHALLENGES

As the metaverse becomes more prevalent, ensuring security within this virtual environment becomes crucial. Here are some key security functions that may be necessary for the metaverse [18, 19]:

1. *Authentication and authorization:* Robust authentication mechanisms will be needed to verify the identities of users accessing the metaverse. This may involve multi-factor authentication, biometrics, or other advanced methods. Authorization mechanisms will control what actions and resources each user is allowed to access within the metaverse.
2. *Data privacy and encryption:* With the metaverse potentially hosting vast amounts of personal and sensitive data, strong data privacy measures will be essential. Encryption protocols can help protect data during transmission and storage, ensuring that unauthorized individuals cannot access or manipulate it.
3. *Virtual asset security:* In the metaverse, users may have virtual assets such as virtual currency, virtual property, or unique digital items. Security measures will be needed to protect these assets from theft, fraud, or unauthorized duplication. Blockchain technology, which provides a decentralized and tamper-resistant ledger, could be employed to enhance asset security.
4. *Anti-cheating measures:* As in any virtual environment, cheating and hacking will be concerns in the metaverse. To maintain fair play and prevent exploitation, robust anti-cheating measures should be implemented. These may include cheat detection algorithms, real-time monitoring, and penalties for offenders.

5. *Content filtering and moderation:* The metaverse will likely host user-generated content, which raises the need for content filtering and moderation. Systems will be required to identify and block inappropriate, harmful, or illegal content, ensuring a safe and welcoming environment for all users.
6. *Virtual identity protection:* Within the metaverse, users may create and manage virtual identities or avatars. Ensuring the security and privacy of these virtual identities will be essential. Measures such as pseudonymity, controlled disclosure of personal information, and protection against identity theft will help maintain user trust.
7. *Network security:* The underlying network infrastructure supporting the metaverse must be secure to prevent unauthorized access, denial-of-service attacks, or other network-based threats. Robust firewalls, intrusion detection systems, and traffic encryption can help safeguard the metaverse network.
8. *Incident response and recovery:* Despite security measures, incidents may still occur within the metaverse. An effective incident response and recovery plan should be in place to mitigate the impact of security breaches, identify vulnerabilities, and restore normal operations as quickly as possible.

10.8   DISCUSSION

As the Metaverse continues to evolve and grow, the development of Metaverse applications is becoming increasingly important. In this chapter, we have explored various technologies and tools that can be used to develop immersive Metaverse applications, including game engines, virtual and augmented reality technologies, blockchain technology, decentralized platforms, artificial intelligence, and decentralized autonomous organizations.

The integration of these technologies and tools can lead to a more immersive and personalized virtual world experience for users, which can lead to greater user engagement and retention. For example, the use of game engines like Unity and Unreal can enable the creation of 3D environments and objects that are visually appealing and responsive to user actions. The integration of virtual and augmented reality technologies can provide a more immersive experience, while the use of blockchain technology and smart contracts can enable user ownership of virtual assets and transactions.

Decentralized platforms like Decentraland and Somnium Space can provide interconnected virtual worlds, creating a more dynamic and sustainable Metaverse ecosystem. The use of decentralized autonomous organizations (DAOs) can also provide fair and responsible governance within virtual worlds, ensuring that users have a voice in decision-making processes. Artificial intelligence can be used to create intelligent NPCs, providing a more realistic and personalized experience for users. This can lead to greater user engagement and retention, as users are more likely to be engaged with virtual worlds that feature intelligent NPCs.

The integration of these technologies and tools can lead to the creation of a fully immersive and decentralized virtual world, providing a more engaging and personalized experience for users. As the Metaverse continues to grow, the development of Metaverse applications will be essential to create a sustainable and vibrant virtual world ecosystem.

10.9 CONCLUSION

The future system for Metaverse application development involves the integration of various technologies and tools to create a fully immersive and decentralized virtual world. Game engines like Unity and Unreal are used to create 3D environments and objects, while virtual and augmented reality technologies are integrated for a more immersive experience. Blockchain technology and smart contracts are used for user ownership of virtual assets and transactions, and decentralized platforms like Decentraland and Somnium Space are incorporated for interconnected virtual worlds. To ensure fair and responsible governance, decentralized autonomous organizations (DAOs) are used, while artificial intelligence is used to create intelligent NPCs for a more realistic and personalized experience. The overall goal of the proposed system is to provide a more engaging and personalized virtual world experience for users, leading to greater user retention and a more vibrant and sustainable virtual world ecosystem.

REFERENCES

1. Castronova, E. (2005). *Synthetic Worlds: The Business and Culture of Online Games.* University of Chicago Press.
2. Chen, J., & Chua, Y. L. (2021). "Blockchain technology for game development: A systematic review." *IEEE Transactions on Games*, 13(2), 102–118.
3. Duan, Y., Luo, X., Zhang, X., & Huang, G. Q. (2021). "Decentralized autonomous organizations in blockchain-based supply chain management: A systematic literature review." *Journal of Cleaner Production*, 314, 127919.
4. Kipper, G., & Rampolla, J. (2018). *Virtual reality and augmented reality: Myths and realities.* A K Peters/CRC Press.
5. Chen, T., Zhou, H., Yang, H., & Liu, S. (2022). "A review of research on metaverse defining taxonomy and adaptive architecture." *2022 5th International Conference on Pattern Recognition and Artificial Intelligence (PRAI), Chengdu, China* (pp. 960–965). doi: 10.1109/PRAI55851.2022.9904076
6. Miao, F., Liu, Y., & Xu, Z. (2021). "An overview of blockchain-based decentralized virtual worlds." *Journal of Network and Computer Applications*, 188, 103074.
7. Rodríguez-Fernández, V., Torrecilla-Salinas, C. J., & de la Hoz, E. (2021). "Artificial intelligence in video games: A systematic review." *Engineering Applications of Artificial Intelligence*, 101, 104312.
8. Johnson, Merrill L. (2022). *Social Virtual Worlds and Their Places: A Geographer's Guide.* Springer.
9. Zhu, Y., Zeng, Y., Li, X., & Wang, H. (2021). "A survey on game engines for virtual reality." *Virtual Reality & Intelligent Hardware*, 3(5), 409–25.
10. Lee, Lik-Hang, et al. (2021). "All one needs to know about metaverse: A complete survey on technological singularity, virtual ecosystem, and research agenda." *arXiv:2110.05352.*
11. Kamińska, D., et al. (2019). "Virtual Reality and Its Applications in Education: Survey." *Information*, 10(318). https://doi.org/10.3390/info10100318.
12. Azuma, R. T. (1997). "A survey of augmented reality." *Presence: Teleoperators and Virtual Environments*, 6(4), 355–385.
13. Srilatha, D., & Thillaiarasu, N. (2023). "Implementation of Intrusion detection and prevention with deep learning in cloud computing." *Journal of Information Technology Management* 15. Special Issue: 1–18.

14. Shyamambika, N., & Thillaiarasu, N. (2016). "A survey on acquiring integrity of shared data with effective user termination in the cloud." *2016 10th International Conference on Intelligent Systems and Control (ISCO)*. IEEE.

15. Thillaiarasu, N., & ChenthurPandian, S. (2016). "Enforcing security and privacy over multi-cloud framework using assessment techniques." *2016 10th International Conference on Intelligent Systems and Control (ISCO)*. IEEE.

16. Preethi, P., Asokan, R., Thillaiarasu, N., & Saravanan, T. (2021). "An effective digit recognition model using enhanced convolutional neural network based chaotic grey wolf optimization." *Journal of Intelligent & Fuzzy Systems*, 41(2), 3727–3737.

17. Thillaiarasu, N., Pandian, S. C., Vijayakumar, V., Prabaharan, S., Ravi, L., & Subramaniyaswamy, V. (2021). "Designing a trivial information relaying scheme for assuring safety in mobile cloud computing environment." *Wireless Networks*, 27, 5477–5490.

18. Gothai, E., Muthukumaran, V., Valarmathi, K., Sathishkumar, V. E., Thillaiarasu, N., & Karthikeyan, P. (2022). "Map-reduce based distance weighted k-nearest neighbor machine learning algorithm for big data applications." *Scalable Computing: Practice and Experience*, 23(4), 129–145.

19. Kaladevi, A. C., Saravanakumar, R., Veena, K., Muthukumaran, V., Thillaiarasu, N., & Kumar, S. S. (2022). "Data analytics on eco-conditional factors affecting speech recognition rate of modern interaction systems." *Journal of Mobile Multimedia*, 18, 1153–1176.

Index

Printed in the United States
by Baker & Taylor Publisher Services